DIS

HISTORY LESSONS

HISTORY LESSONS

A MEMOIR OF MADNESS, MEMORY, AND THE BRAIN

CLIFTON CRAIS

THE OVERLOOK PRESS
NEW YORK, NY

This edition first published in hardcover in the United States in 2014 by
The Overlook Press, Peter Mayer Publishers, Inc.

141 Wooster Street
New York, NY 10012
www.overlookpress.com
For bulk and special sales, please contact sales@overlookny.com,
or write us at the above address.

Copyright © 2014 by Clifton Crais

All rights reserved. No part of this publication may be reproduced or
transmitted in any form or by any means, electronic or mechanical, including
photocopy, recording, or any information storage and retrieval system now
known or to be invented, without permission in writing from the publisher,
except by a reviewer who wishes to quote brief passages in connection with a
review written for inclusion in a magazine, newspaper, or broadcast.

Cataloging-in-Publication Data is available from the Library of Congress

Book design and typeformatting by Bernard Schleifer

Manufactured in the United States of America
ISBN 978-1-4683-0368-1
1 3 5 7 9 10 8 6 4 2
FIRST EDITION

CONTENTS

"Who is it that can tell me who I am?"

—*King Lear*

HISTORY LESSONS

PROLOGUE

—⁕—

I IMAGINE IT HAPPENING LIKE THIS. I WILL BE WALKING along Chestnut Street in New Orleans, watching the squirrels running over a cat's cradle of wires and down creosote poles that stand askew. Or I will be talking on the phone with a sibling and an offhanded remark will spark a momentary flash. Or I'll be reading the newspaper, and my eyes will fall upon an article about unctuous food and jubilant music, or some political shenanigans by corrupt officials. Or the television will be showing a hurricane menacing the always menaced "Big Easy." By one of these everyday occurrences, the memory of a childhood lost will rise to the surface. It will be the smell of gardenias in the summer, or the opening of camellias, or something as simple as "How was school today?" The past will begin revealing itself as if a soft sea breeze was gently sweeping the sands from a monumental ruin that's been hidden right beneath my feet.

I am a contradiction. I am a historian who can't remember. I can recall facts about the past as part of my profession; I can stand in a lecture hall and with just a few notes summon information about a war or what people were thinking centuries ago, or describe complicated historical debates. Dates and events and a thousand details come easily. I have spent a lifetime sifting through the records of others, making connections between the lives of ordinary men and women and the forces that so often defeat them, revealing the hidden patterns of our common past.

It's my own life that I can't remember. For years I have kept a crumpled piece of paper with a list of names, dates, and a chronology to help remind me of my past, specifically my New Orleans childhood. I add to it when I discover something new, a simple date and notation to indicate a birth or death or the name of a school. I labor to commit any of this information to memory, but it disappears as soon as I look up, the sands returning to cover what seemed clear a minute earlier. Even now, I struggle with things like names and faces. Classes of ten or more students leave me utterly bewildered, no matter what mnemonic device I deploy. I sometimes momentarily forget my own children's names, or my wife's, which can elicit strange expressions from a nurse or receptionist. Perhaps he's a new addition to the family, I suppose they wonder, or just a little strange. I worry I will one day forget my self.

I live in the murmur of recollection, a second-person existence. Bereft of time, being no one at all, the "I" that is my self becomes an adjective living the fiction of a noun. The simplest of words—"I am"—become nearly impossible to say because they are inseparable from a past that is irreparably lost.

This isn't the kind of forgetfulness created by a blow to the head or some mysterious illness. It isn't absentmindedness either, or old age's synaptic attrition, though both I am sure will one day play their parts. Every one of us will lose our memory in some way that will trouble our very being. It might take place in an instant—a headache or flash before the eyes. Or forgetting will seep slowly into our minds. There will be times when we will know our histories are running away from us. Unable to tell the story of our selves, we will become absent, trapped not by our lives' silences but by the present and its stubborn indifference to time's relentless passing.

Scientists have a name for my forgetfulness—they have a name for everything, their way of reducing the complex to something a bit more manageable—"chronic childhood amnesia." A mere neurological condition, it is simply the inability to remember one's early years, though this blanket term does little to address the broader question of what constitutes "childhood" and "forgetting." Chronic childhood amnesia is the world's most common form of amnesia and perhaps the least understood. Scientists tend to focus on the most severe causes of memory loss that most often afflict adults: strokes, tumors, certain diseases like Alzheimer's, and serious injuries to the brain resulting from accidents or surgeons' scalpels. My memory problems, however, fall under the heading of "functional amnesias" that result from mental trauma during childhood—circumstances in which life's vicissitudes wound the young brain or somehow shape its development. And by "wound" I mean more than the metaphor "childhood scars" we use to describe trauma and mental anguish. "Scarred," we say. "He was scarred by what happened," somehow marked forever by an experience no child should have witnessed. We now know that behind this image of ineradicable wounding lies both a literal and a figurative truth. Tissues in the temporal lobe may scar as neural connections grow and wither and die as the brain's memory systems develop, leaving childhood memory fragmented or lost.

Advances in the science of memory suggest how this may have come about inside my head. Researchers have focused on two small structures in a region located in the middle of our brains, just above our ears, as especially important to memory's creation and our ability to tell the story of our selves. Neuroscientists refer to this area of the brain as the "limbic system," which is unique to mammals. ("Limbic" comes from the Latin word *limbus*, which means "rim" or "border." Dante used it for the first circle in the *Inferno*.) The first structure, the amygdala, helps regulate automatic reflexes and feelings of fear and aversion, storing stressful events with clarity, telling our brains never to forget—history hardwired. People whose amygdalae have been destroyed have no sense of fear, no matter the threat. The cold blade of a robber's knife pressed across their neck elicits no fear response, no panic, no rush of hormones rousing them to run or fight.

Just behind the amygdala is the hippocampus, which creates and "holds" memories for a while before sending them elsewhere within the brain, in a process that scientists describe as "memory consolidation." The hippocampi play a central role in the creation of declarative and autobiographical memory, upon which we create the notion of our selves through the remembrance of life's minutiae: the taste of bouillabaisse one afternoon in southern France, a lover's smell, a child walking away to kindergarten. They account for the uncanny way a smell or a taste awakens lost time. This capacity to reflect on and organize experience in space and time—to recall the past, tell stories, make associations, create histories—is our brain's most recently evolved memory system. Without it there would be no history, no art and literature, no civilization.

Scientists suspect that repeated overwhelming experiences may have a particular impact on the young brain, which undergoes significant changes up through puberty as structures mature and neurons make connections, or synapses. One part of neural development is the depositing of a fatty material called myelin along the axon (also known as nerve fiber), which acts as a kind of insulation and helps stabilize or complete neural development. Amnesia seems to emerge as a result of incomplete or absent myelination. In children who have been systematically neglected or subjected to repeated trauma, cells die and the hippocampi may atrophy, affecting their ability to bring language and a sense of time to experience and leading to problems with summoning personal memory. These children carry within them a past they do not know consciously but which nonetheless possesses them.

Amnesia is not only the inability to remember—the past that has somehow strayed into the realm of the forgotten. It is the habitation of loss, a life lived amidst silence. For someone with chronic childhood amnesia, it can seem as if the self wanders through the ruins of a destroyed city, unable to piece together exactly what happened, unable to quite figure out why he or she is even there. Life-threatening events may return willy-nilly amidst the oblivion. Memory arrives in fragments—a thousand little pieces of scattered recollection that can't be organized, except for those memories that seem as if they happened yesterday.

My own memory problems began to haunt me more acutely in August 2005, when Hurricane Katrina destroyed much of New Orleans. The city hadn't had a bad hurricane for four decades, since Betsy slammed the city in 1965. Southerners, of course, expect weather forecasters to announce depressions forming off West Africa's coast in the late summer months. A few hurricanes will emerge, one or two of which might hit the United States just as children are bemoaning the start of another school year. But usually life will return to its expected rhythms and cooler temperatures will soon arrive, a harbinger of serene autumn days.

Some people fret about the weather; New Orleanians turn to revelry. The below-sea-level city, with its ring of levees and stations pumping water back into Lake Pontchartrain, likes to thumb its nose at Mother Nature. Booze is the first thing to disappear from the shelves before big storms, followed by other vital necessities such as ice, water, and food. There are bars to hang out in, special drinks to guzzle, the washboard sounds of Cajun zydeco, raucous jazz, and summer's boiled crabs, crayfish, and shrimp downed with ice-cold beer and watermelon sprinkled with salt. Some storms have been bad, but for the most part the city has endured them and emerged the morning after ready for runs to McKenzie's Bakery for sticky buns; to Community Coffee, where the brew drips into tin pots; and, before long, to fried oysters at Casamento's. The camellia still blooms along Saint Charles Avenue, and it's time to plan for another Mardi Gras.

Katrina was different. Around its swollen eye, swirling

gray-black clouds consumed much of the Gulf of Mexico. In the city the light turned milky, then yellow and otherworldly. The clouds looked as hard as anvils and nearly black. Roiling, hungry waters marched inland from the Gulf. Like millions of other people glued to the television, I watched streams of cars fleeing New Orleans. The canals weakened and broke open and waters rushed from the 17th Street Canal all the way to Freret Street, just a long walk from my childhood home. People fled to the Superdome or sought shelter beneath overpasses. There were acts of generosity and heroism, and of extraordinary violence. New Orleans collapsed.

My sister Sabrina packed her Honda before the storm with food and clothes and a menagerie of pets: a golden retriever wagging its tail in excitement, three cats, and a bird. Another cat had gone missing. She put out food and water for it before heading north to Baton Rouge to her daughter's apartment. She figured she would return to a refrigerator of rotten food, and at worst a fallen magnolia in the backyard and a few shingles pried off the roof. A few days later her husband reached the house by boat.

Sabrina had raised a family just a few blocks from the 17th Street Canal, trading the rough-and-tumble Uptown of her childhood for the safety of suburban Lakeview, though she felt like an exile living among the nondescript houses that sat far from her favorite city jaunts. She remained at heart a New Orleans hippie, joyous and disorganized, wishing she could bicycle around town or hang out listening to music while cracking open some crabs, rushing her kids late to school every morning, partying into the evening, living day to day. She and her husband had come into a little money, and after paying off their mortgage, they'd figured they no longer needed insurance. Sabrina lost everything, the roof of her house peaking above fourteen feet of water. She spent the next two years in a trailer, coughing from the mold and the formaldehyde and surrounded by abandoned houses and concrete slabs washed clean by the great storm.

In those months following Katrina, I found my thoughts turning constantly to New Orleans, to a house and a neighborhood, to the French Quarter, to a past I could feel as if surrounded by kindred ghosts: uncles, aunts, siblings, cousins, nephews, and especially my mother. Somehow a very public trauma—entire neighborhoods disappearing beneath the storm waters, the anguished cries of people stranded on roofs, the exiled and disavowed—tugged me homeward. I realized I had spent my life looking to a home and to a childhood I couldn't quite remember, wondering just who I had become—a professor of history, a student of the past—and how I had managed to make a life against the white pages of forgetting.

Memory is notoriously difficult to study, childhood memory doubly so. In some basic ways childhood is lost to us all—in the forgetting that memory demands, in all the remembrances that turn through telling and retelling into family legend—persisting only in the myriad traces that appear in our minds like so many apparitions. Forgetting is a necessary condition of being. So too are the facts we polish into metaphors and the stories we hone into history lessons, processes and results that shape how we live our lives. We wonder what's true and what's lost, what really happened and why.

I decided I had to go back, to try to somehow discover who I was then, what really happened, and what I have become. I had to go searching for clues about my life, become a detective investigating my self. I would walk through the city, eat its food, sit in bars watching people drink, revel in the music all around me. I would abide by the rigors of my discipline. I would interview relatives and neighbors, wade through records, read obscure tomes, take copious notes, scrutinize evidence, test hypotheses, and follow leads wherever they took me. I would create an archive, and from that archive a history. Perhaps I might awaken the past. Memory would return. I could begin putting the pieces together for the very first time. If necessary I would become a memory thief, a historian pursuing his own history, transforming what I found into a facsimile of remembrance. I wanted to know the nature of forgetting. Somewhere within the turmoil of past relationships and even in the folds of my cerebral tissues, I could feel the magical, sometimes awful tugging of a lost New Orleans childhood and the histories bequeathed there from generation to generation.

History Lessons is the story of that journey into the past, a story of loss and silence, survival and recovery, of how we make our way in the world and how that world resides in our very bodies. It is also a journey into the nature of memory and history, science and literature, into our ceaseless longing to know what it means to be human. But above all it is a personal journey. I wanted someone to finally tell me who I am.

I knew I would need to place the data I gathered into some kind of rough chronological order before I could begin figuring out what it meant and where the silences were, chief among them the origins of my amnesia. I decided I was going to begin with the beginning. I wanted to know about my birth and the early years of my life, my parent's divorce and the two deathly experiences I had before turning five. The first was when my mother tried to kill me. The second, a year later in the fall of 1964, was when she tried to kill herself.

ONE

—⚓—

NOWHERE MAN

MY MOTHER LOOKS AT ME AND THEN LOOKS AWAY, at nothing in particular.

"Son, I can't remember. Ever since that operation, well . . . Son, I just can't seem to remember much."

Her memory has simply disappeared, wiped clean.

"You see, son, that doctor must've . . ." and she winds out a story about how things are just different now, how a trauma to the brain or a surgeon's scalpel excised her past.

Nearly two decades ago, my mother suffered a brain aneurysm. An artery ballooning inside her head pressed against fragile cerebral tissues, creating a screaming headache and disorientation. Mom knew she was losing herself. She remembers that it seemed to take forever to walk to the front of her shotgun home just off Magazine Street, three disheveled rooms one after another, to pick up the old black rotary phone and dial zero. She felt her life running away. She tried speaking, but the words would not come out.

The operator traced the call, alerted the emergency services. The medics found my mother unconscious. An ambulance rushed her down Magazine Street and along a warren of side roads to Charity Hospital in downtown New Orleans. A talented neurosurgeon specializing in aneurysms happened to be on call. The artery had ruptured, each heartbeat gushing blood across her brain. A CAT scan revealed the torn artery's location. Quickly, nurses shaved the left side of the scalp. Drills and a whirring of surgical instruments removed a portion of the cranium. Gently repositioning her brain, the surgeon exposed the bleeding artery, around which he placed a shiny titanium clip, much as one might repair a hose.

In the United States, more than 750,000 people suffer a stroke each year, the third leading cause of death. Bleeding inside the brain very often results in serious debilitation. Pools of blood pressing against the brain or cells deprived of oxygen can destroy vital tissues. Victims may lose feeling and control over muscles and limbs or the ability to communicate. Depending on where in the brain the damage occurs, patients often can't remember aspects of their past or solve simple problems that once had come to them effortlessly. Sometimes they lose all the knowledge that once was theirs. They lose their selves. These problems can be fleeting or persistent, and trying to understand them has become the subject of enormous amounts of research. There is no single area of our brain where memory is stored, no cerebral filing cabinet that holds our lives' experiences or the bodily skills we take for granted, like walking or opening a bottle or recalling a child's birthday. Memory entails a complicated interplay among various regions of the brain. Since individual neurons are capable of limited change and any given task typically involves more than one area of the brain, stroke survivors may relearn how to to walk, or memories may somehow begin returning.

Mom was lucky. Most people with the grade of aneurysm she suffered die within days or even hours. Her recovery was uncomplicated, though prolonged by decades of abusing her body with booze and cigarettes and by a lifetime of poverty and unhappiness, barely softened by the recent arrival of Social Security checks. But she came through and, remarkably, stopped getting drunk, at least on a very regular basis.

—⁓⁓—

Only one year before Katrina hit, Mom moved to Florida, and she hates it. She detests its dizzying yet monotonous spread of subdivisions and shopping malls, and the way everyone depends on cars. There's no culture, no personality, she complains, unlike New Orleans, where she walked to fetch her groceries or rode the Magazine Street bus downtown past jewel-toned Victorian houses, as festooned as king cakes, and telephone wires draped with last year's Mardi Gras beads.

She moved extremely reluctantly, after her landlord died and his daughter decided to renovate the shotgun. The house was a wreck. The roof leaked tea-colored water; there was no air-conditioning except for one window unit, covered in dust and mildew and about to tip into the alley; and the entire place was filled with roaches and fleas and with rodents held more or less in check by a couple of alley cats. My mother couldn't find a new place with a rent cheaper than one hundred dollars a month, and that, combined with fears about crime, convinced her to move east.

Mom has two dutiful daughters who live near her. Kinta teaches English in a rough inner-city high school and coaches sports she doesn't know how to play. She's the kind of person who will help out no matter what—steady and sometimes so self-effacing you don't know who she is anymore. Her students are amazed, and the boys utterly thrilled, when she tells them she was once a Playboy Bunny. Kinta's been lying to her husband so she can visit Mom three or more times a week, keeping her company and tending to things like the grocery shopping. Marie, seven years older than Kinta, is a nurse married to a real estate agent, counting the days until she can retire after decades of saving everyone but herself, Mom included. She has proclaimed for so many years that Mom is just about to die that it has turned into a family joke.

It's July 2006, and I have gathered my mother and three sisters at Marie's house in Jacksonville, Florida, to dig through their memories for my own use. I have turned a family vacation into a research trip. The kids are playing in the pool. My wife is inside reading a magazine. Tall oaks surround Marie's house, so we are in the shade, but just a few feet away columned light has found its way through the trees. The crab grass is beechen green, though some of it has turned brown, burnt by the sun. It's blistering hot and very still, as if the heat has slowed everything down. My sister Sabrina has just arrived in her beat-up car from New Orleans, a couple of days late, with a jumble of clothes in a suitcase and a few joints to smoke during evening walks along the beach. Sabrina laughs away her lateness and all her other eccentricities as the peculiarities of being a New

Orleanian, as if they are a species of human that operates by an entirely different conception of time.

I've been traveling around the South for five years now, conducting research and interviews, collecting material, dealing with bureaucracies and their records, ruminating on what I've found and what's not there. This is what most historians do. We spend a lot of time looking for stuff so that we can somehow sketch a picture of people and their world. Historians collect evidence—facts—which they assemble into a story that is also an argument: this is how things happened, for these reasons; this is what this person did and why, and these are the consequences. Historians are more judges than lawyers, constantly contending with conflicting material and theories about the past, hoping that somehow meaning will emerge and, with it, a sense of how things might have really been.

I know I want something more. I am hoping evidence might awaken the past, or at least that knowledge might fill in the spaces where there is now only nothingness. My mother looks nervous, scared even. I am scared too, afraid she will tell me something I actually don't want to know. I try to remain calm and use the steady, dispassionate language my training as a historian has taught me. I'll be a social scientist pursuing factual evidence, determinedly objective, though at the back of my mind I realize I have no idea where this journey into memory and history will lead me. I've packed a briefcase with too many pens and pencils, pads of paper, a few gadgets, hoping the paraphernalia might somehow protect me. Our conversation is stored on a digital recorder smaller than my wallet.

I look at my mother and can tell she once was a beautiful woman. Her eyes, though, seem worn, pained. They are dark and foreboding, and I feel as if in an instant I might tumble into them. I realize I have always lived with the terrifying eyes of my mother.

She looks insane. I want to flee.

Children splash each other in the pool, run around and jump onto rafts.

"Christine. Sweetpea. Please don't run. You'll slip." My daughter looks at me and smiles as she slides into the water.

I want to ask: "What do you remember of our earliest relationship, you know, how we were as mother and young child?" I want to know about one of my few abiding memories, from age four-and-a-half, when my mother tried to commit suicide. It was November, only a few weeks after neighborhood children dressed as ghosts and demons had clamored up our stairs looking for trick-or-treats. The weather had turned cool, finally, and the camellias had begun their winter bloom. She locked herself in the bathroom, laid a towel and pillow carefully across the black-and-white basket-weave tiles, and turned on the gas to the ceramic heater.

The bathroom was opposite the place where I slept. It was more of a closet than a room, with roaches that came out in the night and left welts around my ankles, their fecal matter and the apartment's accumulated debris contributing to a history of childhood bronchitis. My brother Gus, a short, reed-thin boy of fifteen, banged on the door until he broke the lock and it crashed open. And there was my mother lying on the floor.

I don't tell my mother that I have returned to that room. Property deeds disclosed an owner. A letter followed, then a phone call, finally a knock on the door, and an agreement that I would not take any photographs, or use the current owner's name in print.

"My wife thinks you might be some kind of weirdo," Mr. G., the new owner, told me matter-of-factly. "You might, like, come back to rob us or something. We've got a kid, a boy. He's just started school."

My business card with my university and professional title stamped across its front held little persuasive power. But then New Orleans is a rough town filled with scam artists. The city is notoriously violent, with one of the country's worst murder rates. Most everyone cloaks their selves in some sort of protection: alarms, burglar bars, guns, Rottweilers, private security companies, FOB decals on their cars. A curtain of iron seven feet high guards the house. Two "Stanley Security Solutions" signs stand sentry, but Mr. G. worries nonetheless.

A drug dealer owned the house before it went into foreclosure. Mr. G. bought it cheap. Now he's fixing up the house, cleansing the place of its checkered past. The neighborhood is on the upswing. Katrina made Uptown attractive again. Many middle-class whites had abandoned the city for the outlying suburbs, complaining about all sorts of urban menaces: declining schools, political corruption, violence, the disappearance of an urban civility that had once governed the city's chiaroscuro neighborhoods, or at least as so many whites like to believe since civility was enforced by white supremacy. Many moved to the most vulnerable areas, like Lakeview. When the levees failed during Katrina, the waters reached nine or more feet. Entire suburbs disappeared. Now everyone's looking for high ground, and the city is once again a draw.

I walked through the rooms, all smaller than I had imagined. Mr. G. opened the door to my bedroom, which has returned to being a closet and storeroom. My hands reached from one wall to the other. I turned around and there was the bathroom. I stood just as I had nearly half a century ago. Almost nothing has changed. The tile floor is still there with its labored weave of black and white. The shelves opposite the door are still there, which once held hydrogen peroxide, mercurochrome, and assorted makeup. So also the sink, then the toilet, finally the three-quarter shower and bath against the wall. What I saw and what I remembered were seemingly in perfect accord. It felt as if there was no then or now, or as if the experience had become so burned into my mind that it had refused any forces of erosion or dissolution.

I didn't tell Mr. G. what unfolded that autumn day, about all the sadness that could not be contained within that room, or about how the blinding light of memory refuses time's passing.

I pointed to the wall opposite the sink.

"Wasn't there a ceramic heater, about there?"

"Yeah, we removed it. They're too dangerous. Anyway, we got central air now."

———※———

"So, Mom, I want to know more about your life. And 5225 Chestnut Street." What I want to say is, "I want to know what happened between us, you and me. And I want to understand what I couldn't then."

But I don't. It seems too confrontational, too needy and judgmental, as if I am demanding that she offer a detailed accounting of her life. We turn from each other's eyes, and she takes a sip from a tall glass filled with ice and cola. A bead of water, in which a fragment of light has taken a fragile residence, falls to her lap.

"Son, I can't remember."

I know not to press too hard, afraid she will get angry and walk away. But I also know that part of me wants to run from these questions, from our past. I steer the conversation in a different direction, hoping she will reminisce.

I begin by telling what little I know from the research I have gathered so far about the woman who sits opposite me.

"Mom, you grew up during the Depression . . ."

And that's all it takes to get her talking. Despite her protestations, my mother has a remarkable command of her past and can go on for hours until she tires. Perhaps the deficits she complains of relate to short-term memory, to things like what she ate for dinner yesterday, a conversation with a daughter, stories from the books left open and abandoned by her bed. The unbounded ordinariness of everyday life runs away from her, leaving behind a more distant past that seems nearby, as if the 1920s were yesterday and the nineteenth-century history of her family just last week. Childhood experiences rise before her, including radiant stories she had been told by her mother. A history from long before her birth begins surfacing.

I can almost see Mom's grandfather Joseph, the cotton broker, forming before her glistening eyes, walking from his office on the Mississippi River in the Vieux Carré to a fabled house on the palm-lined Esplanade. The wealthy Creole elite lived here, in homes with porches on each floor and seventeen-foot-ceilinged rooms that let the cool air curl around one's feet. There were long dinners beneath gilded chandeliers: oysters from Plaquemines, turtle soup with sherry, a roasted haunch of venison. The adults would drink café au lait in one of the double parlors, adorned with elaborate plaster work, that lay at the foot of sweeping staircases fashioned from mahogany and rosewood imported from Central American jungles.

Just as quickly her mind takes me from the vanished resplendence of Esplanade to an impoverished childhood in a rented shotgun during the Great Depression. Mom lived in Hollygrove, next to the Mississippi River in East New Orleans. It was a poor, predominantly black neighborhood, though the fact that she lived in an African American community is entirely absent in her recollection. From the photographs and records I have examined, Hollygrove was a crisscross of oyster shell roads and ditches, into which a few cars dipped uneasily, so low and underserved by the municipal government that it regularly suffered some of the worst flooding in the city. The summer storms left the houses sitting on their stilts like little islands in a suburban lake.

Mom was eight years old, the youngest of six children, when the stock market collapsed in 1929. At first New Orleanians thought themselves immune from the crash's effects. They'd ride it out like a hurricane or like any other nuisance that rolled into town—celebrate in the face of impending disaster, maybe even conjure up a new cocktail, dish, or ditty to mark the occasion. Wall Street seemed a continent away, financial turmoil a distinctly northern malady. Furthermore, they had more important things to fret over.

The city ran on alcohol, sex, and corruption, and Prohibition had made these even more profitable. Federal agents stepped up their patrols along the coast's Rum Row. Suddenly it became more difficult to smuggle thousands of cases of Cuban booze from the boats sashaying at anchor off the Balise. City authorities also began yet another effort to rid the city of its great "devilment": gambling, prostitution, and burlesque dancing. Some feared the city's citizens would eventually go blind from epidemics of syphilis and the clap. Amidst the endless political shenanigans and charges of corruption, on two things everyone could agree: the "Green Wave," Tulane's football team, was having a great season, and Mardi Gras would continue as ever.

The bottom fell out in the middle of 1930, just in time for Mom's ninth birthday. Jobs disappeared. The municipal government didn't know how to respond, spending more time squabbling with the state and federal authorities than helping its citizens. Unemployed workers rioted. Violence swept the city. Ineptitude and indifference continued. Corrupt officials worked hard to ensure the money from bribes kept flowing. New Orleans would become the largest municipality in the country that did not provide money for the relief of indigent families. The best Mayor Walmsley could do was to hand out oranges for the needy to hawk on street corners. Before long the city was awash in citrus.

Crisis set in by 1931. The river fell quiet. Even the city's prostitutes reduced their fees. The first and hardest hit were men without skills, men like Mom's father, Zeno, who took whatever work he could find. Mostly he dreamed. Mom's mother, Cecile, made a little money teaching English to Italian immigrants. But the family was desperately poor, living by their wits and by gifts of food and clothing. Often they couldn't make the five-dollar rent and took up odd jobs or borrowed from family to make ends meet. The children went hungry more often than not.

Mom listened to her mother's stories of their wealthy past and tended her father's dreams. She scavenged Uptown for Coke and Barq's Root Beer bottles, while in the backyard her father concocted "Zeno's Magic Furniture Polish." Mom spent hours swirling pebbles in the bottles to clean the dirt that clung to the inside. The family would then fill the empty bottles with the polish, which Zeno would take to the Garden District and the wealthier neighborhoods, where he'd sell his magic door to door. He felt more at home wandering through the mausoleums at Lafayette Cemetery, which looked like a fabled city—an ancient Rome or Athens in miniature—than knocking on the doors of stately homes, with their nineteen-foot ceilings, crystal chandeliers, and empurpled plaster-work framing windows. But he had to go where the money was.

"I'm gonna sell the formula to Johnson & Johnson," he told his family, even as bottles of the polish piled up in the backyard. They would be rich. Mom would get her pair of patent leather shoes. The family wouldn't have to scrape by. They would return to their rightful place among the city's elite. They would own a house far from the railroad tracks, a solid house two or even three stories high near Saint Charles Avenue or maybe even in the Garden District. There would be Sunday dinners at Galatoire's, weekend trips to the Gulf Coast. When the weather had turned and winter brought out the camellias in pinks, coral reds, and silken whites, they would attend one of the Carnival balls.

They would throw beads and doubloons from dazzling floats, instead of scampering after them on Mardi Gras day.

Mom lived by her father's promises and her mother's stories of their illustrious Creole past. Mom might have gone to sleep hungry, but she could at least sate herself with family history. "We lived down on," a story would begin. By that simple pronoun—"we"—Mom could escape time into some eternal familial existence, a past that was right there to behold and always would be, a past that would somehow wash clean the stain of poverty and despair. As long tales unwound of how things once were, she would nod yes, as if the stories might, at any moment, become real again. "It will come right. It will come right." This prophecy at the end of a story softened the wounds that gnawed at her like a curse or a sickness in the soul. Mom would have those patent leather shoes. She would meet her prince.

In the middle years of the Depression, Mom blossomed into a young woman with dark, flowing hair and an hour-glass figure. Whenever she had a few coins in her purse, she would walk down the block to the Ashton Theatre on Apple Street—the "Ashcan" everyone called it—a hulking temple of a place where upward of seven hundred people at a time could watch newsreels and dream away the day. The large Reproduco pipe organ from Chicago sat silent. The "talkies" had arrived, and the great films of Ernst Lubitsch, Howard Hawks, John Ford, and Alfred Hitchcock, and femme fatales like Dorothy Lamour, Greta Garbo, and Rita Hayworth —sophisticated women, beautiful, powerfully seductive.

Mom's stories and the material I have collected meld together, and in a confusion of tenses I feel myself tumbling into more distant histories. I look at her again. Time ripples across her weathered face as if an earlier self, formed of impossible longings, still lies just beneath the surface. Mom wanted a mink coat, a swimsuit that held tight to her bosom, a lover who would return her to what the family had irrevocably lost: wealth, stability, good taste, a house she would own, and adoring, successful children. Dreams were her only inheritance. She yearned for a man who would whisk her away. He would be a good man, a decent man, most of all a successful man, who would take her to Easter brunches at Commander's Palace, where the gin fizzes smelled of spring blossoms and waiters served Oysters Rockefeller upon a plate of diamonds. In the summer he would take her away from the city's swelter and its deadened talk of the weather. In the tender hour after a thunderstorm, he would come to her, and the cricket lullabies and fireflies sparkling in their nightly waltz would tell her that she was content, and secure, and in love.

She tells me how she and my father first met at an evening class at the local public library. It was 1938; half her life had taken place under the Depression. Dad was in his mid-twenties, eight years older than Mom, a handsome man who wore his fedora like Bogart. He owned a car and spent weekends at Audubon Park in his whites, perfecting sliced backhands across the red clay courts. His family lived in a neighborhood of middle-class professionals. Their solid, square house near Claiborne sat high off the road, as if it were not of the city and belonged instead to an America of modest success and restraint. They had done well enough to purchase summer cottages in Monteagle, Tennessee, a southern Chautauqua meant for relaxation, Christian reflection, and learning, nestled high up on the Cumberland

Plateau. For a short while Dad attended Tulane, where the Crais men were expected to earn degrees that would ensure continued familial success, then studied for a semester at Vanderbilt before heading back to New Orleans.

In teasing out details scattered across various archives and books, I discovered that my paternal family's economic success and Christian footing belies a more complicated past. Suddenly I find myself researching Eastern European and Caribbean history, somehow hoping that these pasts might begin revealing the worlds that created my parents and, by a simple fact of biology, me. Genealogical work is often this way, particularly in an immigrant country like the United States. Most of us come from someplace else. I move up and down the generations and across oceans as if I were playing scales on a piano. I begin with my father's maternal grandfather, Jacob Rosenberg, who emigrated in the 1870s from an area in northeast Poland near the border with Lithuania, a contested territory that passed from one owner to another in the fraught politics of Eastern European nationalism. Jews in particular suffered increasing persecution in the years following the 1863 Polish Uprising. Starvation followed failed nationalism and radical politics. Many perished in the famines of 1868 and 1869. After 1870 Jews began migrating to the United States, South Africa, Sweden, and South America. Those who remained in this small corner of Poland—about 7,000 by 1940—were shot in a nearby forest or murdered in the gas chambers.

A sickly man, Jacob was meant to join his three brothers in South Africa. He made his way from Poland to Italy but ended up somehow in Texas and then in the wet lowlands of Plaquemines Parish, Louisiana, working on the docks or as a drayman. Jacob married into an old French family from Pointe à la Hache. Grace Agnes Martin bore him five children in their short marriage of a dozen years. Jacob wanted them to have Jewish or Polish names, some recognition of his homeland and his faith, the past that had led him to America: they named my father's mother Sarah, after her grandmother, Sara (Chaisora) Lewenshon.

The family had just moved to New Orleans when Jacob died at the age of thirty-five. Sarah Rosenberg was five, her baby brother barely a year old. Little of her father's past remained beyond her name. The pogroms and starvation, Jacob's epic journey as a very young man across Eastern Europe to Italy, Yiddish, the Torah, his beloved brothers working in South Africa, his family back in Poland—all this history faded away. Sarah grew up with her mother in New Orleans, raised in the Catholic Church until 1905, when a request for money to bury Grace led the indignant eighteen-year-old Sara down the street to the Presbyterians. Two years later, Sarah married Gustave Henry Crais. They moved into a small shotgun with Sarah's younger brother.

Sarah's husband, I learn, also had forsaken Jewish roots. Gustave's mother descended from the Ashkenazi Jewish community of Alsace and western Germany, who immigrated to the United States in the second decade of the nineteenth century: Mayers and Salzmans, Koenigs and Schocklers. All this Eastern European and Jewish past has been forgotten, as if a branch of the family tree was lopped off or never even existed. Male descent is what ended up counting, and that means Louisiana's Catholicism and its distinctive French past. But even here there are all sorts of strange silences and elisions of time and space. "We are French," we like to say, "Creole." The Crais family came from France. True enough, as long as one forgets Haiti.

I return to my genealogical spelunking, searching for records scattered halfway round the world: Aix-en-Provence, the Caribbean, and across the United States. The key is to work backward from the present to ever deeper pasts, and then, once you've found something, to research horizontally, situating the person in their time and place as if you were an anthropologist conducting fieldwork. Establishing a context is one of the most important challenges of all historical work. Slowly you begin dreaming up complex social worlds: the clothes people wore, what they ate, their labor—in short, their lives and deaths. From there you can begin working toward the present again, which in my case means finding out who my grandparents were, how my parents met, and wondering how these more distant pasts might have shaped later generations.

The original American immigrant, Joseph Lange Crais, sailed to New Orleans as a young exile, among the last whites to flee Saint-Domingue's bloody slave revolution during the final years of the eighteenth century. I can identify almost the day he arrived in New Orleans. He was a Haitian creole, not an American one. I start reading more deeply into the island's turbulent history, push further back in time. Joseph's father had left eastern France in the second half of the eighteenth century to make his fortune in the insufferably hot island, ending up on the northwestern tip of the island in Le Grand Dérangement, the "Great Madness," when the British prosecuted and scattered Acadians and other French colonists across the Americas.

Môle-Saint-Nicolas was a dreadful spot. Over half of the seven hundred people who first landed there perished within a year, the "air putrid with the dead." Joseph's father became a *petit blanc*, a "small white," most likely a ship builder, successful enough to own a modest house. At the time Saint-Domingue was France's most valuable North American possession, by the end of the century the world's largest producer of sugar and coffee, thanks in no small part to the 160,000 slaves toiling on its plantations. The island had the most brutal system of slavery anywhere in the New World: laborers worked to death digging canals and cutting cane, arms torn off by the mill, fugitives burned alive in Le Cap's public square. Slaves resisted, some escaping to the remote jungles of the interior. In 1791, a year before Joseph's birth, they rebelled. Inspired by the French Revolution and memories of ancestral homelands, slave armies attacked Le Cap. A thousand plantations burned to the ground. Two thousand whites died. Upwards of ten thousand slave rebels died, their leaders' heads mounted on stakes in the main square. The Black Jacobin Toussaint Louverture emerged as the new leader, declaring in 1793 that he would accept nothing less than emancipation for the island, while putting down revolts in the north with spectacular violence. Napoleon's attempt to restore white rule in 1802–3 led to a vicious war. Louverture died in 1803, his soul returning to lead an army of African dead. In 1804, when Jean-Jacques Dessalines proclaimed Haiti's independence, black revolutionaries massacred many of the remaining white planters in a spasm of racial violence and retribution.

The revolution sent waves of exiles into the Caribbean and the United States. Joseph was one of nearly 20,000 people who left the sugar island for the lower Mississippi River, a group that included not only whites but also slaves and especially free people of color. The migration nearly doubled New Orleans's population. In his early thirties Joseph married another exile, Emilie Doucet, nearly a decade his senior. He became a shipwright during a time of explosive growth in the river traffic moving up and down the Mississippi. The refugee couple thrived in their new home. Joseph brought into his marriage at least one slave; by 1840 the couple owned ten slaves, a considerable number for an urban family. A decade later, when the household included four mulatto children, Joseph owned real estate valued at $10,000, today a quarter million dollars. His twenty-eight-year-old son, my great-great grandfather, owned property worth $3,000 and a house on Bourbon Street, and worked as a "professor."

My father's father began work as a clerk in an insurance agency at the young age of fourteen. He became the head of the fire insurance department at Black, Rogers and Company, where he worked for twenty-two years until his death in 1939. By this time, memory of these immigrant pasts had long since disappeared within the making and remaking of familial history, as if their kind of American success demanded a certain silencing. A stolid, white, Christian, middle-class life of quiet accomplishment could not contain histories of European Jewry, Caribbean slavery, and exile. In the decades following the Civil War, Haiti came to be seen as too marginal, too poor, and most of all too black, in a period also associated with rising anti-Semitism. In any event, when my parents met in the late 1930s, Dad was a racist and an anti-Semite.

Dad's parents looked askance at this young woman entering their errant son's life. They hoped he would return to college, marry someone of similar respectability, and make something of himself. My mother and father dated through the summer heat and humidity and spent Labor Day of 1939 on the coast, alone and happy. Dad's father died two months later, and two months after that, the couple married. They soon left for New York City. Mom was nineteen. She got her coat, not a mink but at least one with a fur collar.

Dad enlisted in the Merchant Marines after the birth of their first child, Susan, in 1941. Mom loved New York: the change of seasons, the first snow, the city's hustle and bustle. The Merchant Marines brought the couple a regular paycheck—the first time in Mom's life she could make ends meet, though with it came the dreadful fear that a German torpedo would sink her husband's convoy somewhere in the North Atlantic. The end of the war returned them to New Orleans, and after that, children began to arrive every few years: Marie in 1945, my brother Gus four years later, the twins Catherine and Mary born prematurely and dead from holes in their hearts, followed by three children amidst a collapsing marriage: Kinta in 1952, Sabrina in 1955. I was born five years later.

Across Louisiana and Texas the petrochemical industry was booming. Refineries converted millions of gallons of oil into gasoline for America's automobile revolution. After a short time in New Orleans, the family moved to a dreaded suburb outside Houston, where rattlesnakes cooled themselves under the house, then moved again to the Mississippi

Gulf Coast, where low tide waters raced to the horizon. At least until Houston Mom tried hiding her frustrations over the constant moves and living month to month. Dad was good with numbers, and jobs were easily had, but he could never keep one for much longer than a year. He argued with his employers, complained about his coworkers. Nothing was ever good enough, and a better job was always around the corner. He also borrowed money against his paycheck. In the late 1950s the Continental Oil and Transportation Company pursued in him court, securing a judgment for $181.53.

Mom can't name all the different companies Dad worked for, or all the addresses where they lived, though just talking about it still makes her angry. I know that for some time she tried being a dutiful wife and mother. There were cakes to bake, Mardi Gras costumes to sew from white, purple, and green cloth that hung in strips across the old Singer. For days her feet rocked up and down. The pulley whirled until she had made her children into princesses, heroes, jesters. She insisted on celebrating Mardi Gras even in Texas and Mississippi. Neighborhood children dressed up. For a long day, the house they were renting became a carnival ground filled with games, children running to and fro, fried chicken and potato salad, and plenty of sweets.

Back on the porch in Jacksonville, a smile begins to form, and almost as suddenly something happens. Her eyes blink two or three times as if her vision has gone blurry; I wonder if a speck of dirt has tumbled into them. Her mind turns to the period around my birth and young childhood, when I was needy and she was sick.

"You see, son," Mom begins, going on to explain how

hard life was in the final years of her marriage to my father, and how much she despised the way Dad constantly moved the family around as he looked for jobs that were never good enough. My mother craved stability, anything that might take her away from her life's poverty. She wanted a house, family suppers at six, vacations on the Gulf Coast, a middle-class life where she would no longer worry about how to make it month to month. Dad never finished college, never kept a job for long, never earned enough to buy a house. Mom came to think her prince was a coward.

I imagine how difficult being divorced must have been: a housewife for nearly two decades, no qualifications, not even a high school education, a clutch of children. She expected marriage to last forever and to always be cared for, no matter what happened. Divorce was anathema. Louisiana's legal system had very few safeguards for women. Divorce meant abandonment, not least from the Roman Catholic Church.

Words drift away from our relationship to the context within which it unfolded. Mom looks at me, sees my successes: a tenured professor at a prestigious institution, a stable and fulfilling marriage of over twenty years, two wonderful children.

"We all suffer. It does you good, builds character. Look at you, how well you've done. We're all so proud. And she's so beautiful. And those children of yours . . ."

How lucky you are, she tells me, how intelligent and successful. I don't respond; I am not sure I would know how.

My mother looks at me again. This walk through the ruins has wearied her. She returns to where we began, memory's disavowal. I realize her memory seems to unravel around the time of my birth and childhood, a very late child coming into a broken marriage. She has a hard time remembering all the bad stuff—alcoholism, suicide attempts, the divorce, the abuse of her children, the incalculable damage she has wrought on others—but I also wonder if she fears being devoured by the past, by all those unrequited longings and endless disappointments.

—⁂—

"Do you remember that time when I was two or three years old, and you drew a bath in the old enameled cast-iron tub until the water reached nearly halfway up? We always used Ivory Soap. It's one of the things I remember from childhood. I liked the way the soap bobbed up and down and left white trails in the water. For a little while you washed me, until you pushed my head under the water and kept me there. I struggled, kicked, slapped at the water, flailed about helplessly."

I am not sure why I want to ask her this. I know the facts of what happened, but not why. Decades later, I ended up in the arms of a woman, a close friend of one of my sisters, in a cheap New Orleans hotel, and she told me this story about me. All of this now seems as real as a 5mm family film projected against a white sheet in endless loops of yesterday, today, next week. And now I am talking to the person who tried to kill me, wanting to know why, wanting to rail against her. But somehow I can't get the words out, and so it is a secret we both keep.

—⁂—

Water under the bridge, Mom likes to say, let sleeping dogs lie. When we are together, once or maybe twice a year, we are mostly lost to each other, as though we were islands along some archipelago made of family, connected certainly, but also ineluctably separated by our silences. My mother wants to forget. She thanks the doctor for helping in the effort. Forgetting is her defense. It may be her way of life, a mode of existence. But I want to know. I want Mom, or rather the two of us, to create a past, to tell a story of us together, no matter how painful and incomplete. I have conducted research, done other interviews, looked at photographs, examined documents, assembled chronologies, chased sources from France to Haiti and beyond. I know about her family's history, as well as my father's. But I want something more than what documents can tell me, much more. I want her to take me back to the abyss. She is the only one who can do it. Mom is an original source. With the other information I have gathered, I hope I can produce a proximate history of what happened, and a possible reason for my inability to remember.

I wonder if neuroscience might offer some clues, might in some way help me imagine what happened all those years ago and answer the question of how I became me. Researchers have discovered that depressed mothers alter the brains of their babies by the way they look at them, by their very eyes. Experiments have shown that anxious or conflicted early attachments may result in the overproduction of biochemicals that shape neural growth and the biological processes involved in memory, fear, and forgetting. The medial temporal lobe, which scientists have identified as central to the creation and storing of memory, undergoes terrific transformation in childhood and is especially susceptible to the effects of neurotransmitters and other biochemicals.

The child's external world of emergent relations unfolds inside the brain in a kind of biological conversation with genetic inheritance. Certain genes switch on, others turn off. Synaptic connections are made, pared down, remade. Entire structures evolve that shape how we remember, how we tell the story of our selves, of who we really are. These stories—the mental histories we assemble over the course of our lives—become our "selves." We sift through the remains of our being, assigning particular importance to this or that recollection, or discarding pasts we once thought determined who we are. Stories change, often in the telling. In this way, no matter how fragmentary and unreliable, memory functions as the archive from which we produce a history of who we have become. The autobiographies we create shape the way we live our lives.

A distinctive feature of human memory and storytelling is its relation to witnessing—the idea that we are aware of our selves in the world as participant and as observer. Witnessing is central to the self, a fundamental part of being human, and the origin of historical consciousness. It is one thing to have a mind, quite another thing to have self-awareness. This latter quality provides us with the ability to make meaning, to make our very selves in the sensibilities of space and time. Witnessing, the neuroscientist Antonio Damasio suggests, allows us to convert actions into emotions of worth, a hug into affection, pain into suffering. Without witnessing we could not tell a story. Without witnessing there would be no literature, no history.

Memory and narrative in some senses are thus made

at the cellular level. In stressful situations the infant brain produces an excess of cortisol, one of the chemicals that helps prepare the body to "fight or flight," except that beyond crying or making a fuss the human infant is defenseless. Cortisol plays an important role in how we remember. Increases in cortisol help encode and consolidate certain kinds of recollection, especially those relating to punctuated, highly emotionally charged events, creating in the human mind what neuroscientists call "flashbulb memories." My vivid recollection of the bathroom that day in 1964 when I stood as witness to my mother trying to kill herself is one such example.

The crystalline, lucid image of a traumatic experience is not perfectly accurate, though it feels that way. Flashbulb memories are no complete replica within the human mind of what happened. They can be inaccurate, and they exact a certain cost. The very same cortisol that helps create flash-bulb memories also destroys neurons, quite literally obliter-ating memory surrounding the traumatic experience. A likeness exists as if it were a photograph, but shorn of its surround-ings. In my case, I have no awareness of what immediately preceded the event or came later, who may have swept me up in their arms at that difficult moment, or if I stood there alone.

My childhood turmoil unfolded physiologically. The external environment became part of changes taking place inside my head. My inability to remember childhood is ultimately a neurological deficit, but I wonder if the origins of my memory problems lie not only in trauma. There may be an additional reason rooted in the culture of nurturing and its absence. The family's dissolution and Mom's collapse into alcoholism and depression brought an end to storytelling. The familial narratives that are so important to who we are and to how the past is handed from one generation to the next disappeared, including those that remained untold until now. I forgot because so much was being forgotten.

—⁓—

I once believed I could be the story I told about my self. I could speak of experiences like so many stepping-stones. *Reminisci ergo sum*: I remember therefore I am. But what happens when you can't remember, when there are too many blank spaces? Without memory's inscrutable insinuations I am a nowhere man, never certain where I am at any moment or if anything really fits together. In some basic way the past is forever lost. And yet, paradoxically, the past exists within us all, history transmitted from one generation to the next, sometimes plainly, often in whispers. When the poet Rainer Maria Rilke wrote that our ancestors "still exist in us . . . as [a] burden upon our fate, as murmuring blood," he was recognizing the surreptitious ways the past endures within the present. Most of us have had the feeling that a distant past somehow seems as if it were whispering in our ear, momentarily calling us away from the expected motions of life. Time may rend our desire to reach across the blankness, but the hunger to return persists, to somehow be there, someplace different.

I wish that I had relics from my past—books, toys, puzzles, anything really, no matter how tattered—that might awaken memory, help fill in the spaces where my childhood body was. My wife has all sorts of such things, as well as hundreds of photographs. The albums sit above our television. From time to time she looks at them and stories unfold in the neat progression of childhood, adolescence, friends, marriage, children, the deaths of loved ones, our lives together. Her mother has more albums. When the two of them are together, they flip through the photographs and talk about this and that, and the hours melt away.

There exist only a few photographs of me as a child, not more than a half dozen. I've claimed most of them only in my middle age in a kind of rescue effort. I have acquired other photographs from family members, but these don't help much since I can't identify the faces or occasions.

In my professional life, I have a vast collection of evidence from other pasts, four filing cabinets filled with photocopies of documents from archives across the world. I have thousands of notecards full of words a century or two old, hard drives stored with notes and digital photographs of archival records. I make backups, file evidence in manila folders, order and reorder these gathered relics.

Virtually all of this evidence concerns misfortune. Like many of my colleagues I have been drawn to the study of conflict, oppression, suffering, slavery, peonage, landlessness, poverty, death, infant mortality. I have spent a career among the dead, and more precisely among those who died before their time. Historians are drawn to tragedy, though we usually disavow what we find when we go looking into the past. This is our professional conundrum, running to the past and then running away from its horrors. We are students of trauma, of wars, murders, genocide, and human suffering. Ours is a prose both of longing—to witness what has happened—and of mourning—for our inability to be there.

Most historians believe that language contains within it something real. The dead have left words and other artifacts behind. Our job is somehow to make contact with them, to create some sort of relationship with these bits and pieces. Repeating again and again words rediscovered, we imagine being there, rescuing posterity from oblivion. We work amongst ghosts, dreaming we might sit down with them for a while and talk, knowing we can't. From these traces of language that have endured well past the moment they left someone's lips, we can begin reconstructing the past, not as it really was but a probable past, a history that somehow triumphs over death.

We can try to listen in on the past, bear witness to what unfolded, glimpse how people tried to make sense of what was happening to them amidst swirling confusion, learn something about how the past comes to be transmitted from one generation to the next. All the while we work with silences. Of course there is the simple absence of evidence. There is so much that is not there—pasts not recorded, material that time has whittled into nothingness. But I wonder if silences aren't also created within the evidentiary record, as if something exists only by the simultaneous creation of its opposite. Presence and absence, memory and forgetting, remain inextricably intertwined, often in struggle, as in my mother's forgetfulness pitted against my attempts to remember. Historical memory exists only by a radical paring down of sensory experience and the ordering of evidence into a narrative of who we are—our pasts, our dreams—amid the ruinations of human existence. The past is a mess, a bloody terrible mess of infinite horror. We historians spend our professional lives in its viscera, and also in the silence passed from one generation to the next like a hungry wound that is there and then gone.

—◊◊◊—

For a long time I refused the veracity of that first memory, of my mother trying to take her life. I can see her there on the bathroom floor. She wants to die. Even a child could tell something was terribly wrong, though at the time I could not put into words my mother's despair. But it is the knocking on the door that comes to me most vividly.

"I was coming home from school," my sister Sabrina tells me. Kinta, Sabrina, and I have the Jacksonville beach pretty much to ourselves. Mom is back in her apartment, exhausted by the family gathering and by my questioning. Now I am hoping my sisters might share with me their memories of that day nearly a half century ago. A ring of late afternoon thunderstorms a mile or two away has kicked up a breeze and sent everyone home for the day. We huddle near a lifeguard station. I watch the storm, count the seconds between the lightning and the rotund booms racing along the slate black sea. Not more than fifty yards from the shore, dolphins are jumping into the air, smacking their tails against the water, and in between talking about madness and our mother, Sabrina remarks on the beauty of the sea and the wondrousness of dolphins chasing fish.

"Oh look, one jumped right out. That's so fine, so so fine." We follow Sabrina's words out to the sea, always a second too late.

"Yeah." Sabrina takes a swig of beer. When she is on vacation Sabrina is seldom without a beer or glass of wine in her hand and a joint in her purse. I half expect her to suggest we take a swim among the dolphins in the middle of a storm. Kinta marvels at her recklessness. Mom's alcoholism scared Kinta away from booze and drugs; she keeps her distance from our family scourge, finds a certain safety in tentativeness.

"Yeah, I was coming home from school, and Gus was breaking down the door." Sabrina was in fourth grade then, Gus was fifteen, and Kinta three years younger. The girls had walked home together past magnolias and the last hydrangea blossoms, saying good-bye to friends along the way, up to the cement steps, where they had heard the sound of our brother banging his fists against the door until the lock broke. The door crashed open. Someone called Marie, our nineteen-year-old sister, who took Mom to the hospital.

A few weeks later I am driving down to New Orleans with Sabrina, who was left stranded in the Atlanta airport on her way home from visiting Mom in Jacksonville. By chance I had been heading from Atlanta to New Orleans the next day, to visit Sabrina but more importantly to conduct some research, so the two of us have woken early in the morning for the seven-hour drive. Barreling along Highway 10, we drink coffee, gossip about family. We talk about our childhoods. I describe some of the reading I have been doing on attachment and maternal bonds, on how the brain develops as if in some symphonic relationship to the world of which it is a part, and on how we have narrative brains. On the bridges we pass I find myself pausing to look at men in flat boats fishing for sheepshead, redfish, and speckled trout, and a memory arrives, age seven or eight, of dipping a cane pole into the muddy Gulf.

Turning my head for a second, I ask my sister simply, "What is your greatest hurt?"

She replies right away, as if the words were formed long ago, an answer waiting for the question. "Mom didn't love me. I know she didn't love me."

"I don't think she could love anyone then." I talk some more about maternal attachment and depression, but our conversation quiets.

I drop Sabrina off at her father-in-law's country house on Bayou Lacombe. I need to make the next leg of the trip alone. I have a three o'clock appointment at the Southeast Louisiana Hospital, a state psychiatric facility. I turn onto Highway 190 and drive through Mandeville. I glance at my Mapquest directions, see a sign for Fontainebleau State Park, and know I am close. A final left hand turn will put me on a straight road through a pine wood to the hospital.

I'm early, so I wander through the park. The stifling humidity fogs my eyeglasses and camera. A train track has been turned into a bike path. Except for a few rangers, the park is empty. Half a mile in, I come upon the ruins of an old building. It seems industrial; perhaps sugar was produced here when Louisiana plantations grew wealthy by slave labor. I look around for signs but find nothing, just red brick amid oak trees and palmettos.

Directly opposite the entrance to the hospital is a derelict bus stop. Families visiting patients would have gotten on and off here, but the county authorities discontinued the bus service. Nowadays everyone drives a car. No one has sat on the bench for a very long time. Someone will tear the structure down, or more likely it will simply disappear into the undergrowth.

The hospital is mainly used for drug rehabilitation, for children and adolescents, and for people who have gone off their medicines and need to be re-stabilized. Most everyone leaves within two years.

At the entrance to the hospital is a pine tree that has been damaged by Hurricane Katrina; the tip is bent nearly to the ground. During the hurricane, most of the patients had to be evacuated. Quite a few of the buildings suffered considerable damage. County services had to cut down many of the trees, so there is little shade now. Everyone tries to stay inside. Except for the parked cars and hum of the air-conditioning compressors, the campus looks deserted.

"I have a meeting at three o'clock with Ms. Washington."

"Take your first right, go past two roads, then left, then a cement driveway, and Volunteer Services."

I already feel lost, but I nod as if I understand him completely and head down the main road before turning right. I pass a man in a pickup truck wiping his eyes with a large white towel. I take a left at the sign for "Volunteer Services," and park. The windows have bars, behind them darkness. I can't find an open door, so I wander around looking for an entrance until one opens, then I walk down corridors until I meet an attendant. He doesn't know a "Ms. Washington." I am lost. After a few minutes I get back into the car and drive through the maze of single-story buildings. I see another employee who points me in the right direction.

I am writing a book, a book about my life and my mother's, our relationship, I tell Ms. Washington. I hand her my business card. They are not used to this sort of request, a son in search of a particularly bad moment in his mother's life. I know this is peculiar. I worry that she thinks I'm crazy. I worry that I am crazy. I try to appear patient and professional, though my head is spinning, and I am lost in the fear that at any minute I might fall apart, dissolve right in front of her. I'll end up in the same room as my mother all those years ago.

Ms. Washington is actually calm and professional and tries to be as helpful as she can. She explains I cannot interview patients, nor can I use the names of any patients I might come across in the records. No photographing either, though Ms. Washington does assure me that at the end of my visit I can formally request photographs.

I had hoped to find records that would allow me to hear my mother's voice. I had hoped that there might have been someone who had cared to listen, who by their empathy had transcribed my mother's pain onto paper, creating an archive that would give me a little fragment from the past to hold onto, not as a child but as a knowing adult. What had a doctor seen in my mother's eyes? Perhaps not terror but the anguish of her suffering? In fact there are no extant records, no textual trace of the patient "Yvonne Crais." Everything had been microfilmed, the original documents destroyed, but in Louisiana's heat and humidity the chemical residue on the film continued reacting. What once had been clearly legible now has turned to a blank darkness.

Ms. Washington tells me she did locate a few photo albums from the 1950s and early 1960s, around the time hard-drinking and womanizing Earl Long—the "last of the red hot poppas," as he described himself—ran the governorship from the grounds of the mental hospital. I gather some data about gender and race and how long people stayed here: eighteen months for most adults, many years for some with no place else to go.

I look around, imagine what it must have been like for my mother, attempt to be there in the past, a witness again. I try looking through her eyes, those obsidian eyes I fear. I try looking from the vantage point of others, siblings and family, a soon-to-be-ex-husband, doctors and nurses. I think of context, of time and the constraints within which people's lives were made or, in my mother's case, unraveled. The only perspective I cannot imagine is my own. This is perplexing. I have no memory of when my mother was in Mandeville. When I try returning to that time, my mind becomes not blank but a kind of graveled grayness, as if the transmission has simply stopped, or as if a tape, long erased, won't stop playing.

I walk with Ms. Washington to a room where I can examine the albums. I sit down, gather myself. Mental hospitals are frightening places, particularly state institutions. The hospital was segregated by race and gender. There were children as young as five or six, who seem to have spent a good deal of time with adults; I worry that the abuse some of them may have suffered continued within the hospital's confines. The hospital staff provided the patients with many activities, including some, like sewing and carpentry, they might use on the outside. There were billiards and card tables, a large outdoor swimming pool, athletic events, and regularly scheduled dances. A library filled with books. One photograph is of a puppet theater. The show is *The Wizard of Oz*, except that in this case it's the wicked witch who towers over the good witch lying crumpled on the ground. Many of the photographs are of Mardi Gras, when everyone dressed up for the hospital ball presided over by a King and Queen and child attendant.

Ms. Washington tells me later that there are fewer occupational services today than a half-century ago, though the hospital has roughly the same number of patients. Hurricane Katrina strained Louisiana's mental health services to the breaking point. And just when everyone thought things couldn't get worse, the recession of 2008–09 hit. For decades state services have been grossly underfunded. Most patients receive minimal talk therapy. They watch a lot of television.

Mom entered the mental health system at a time when hospital conditions diverged dramatically and treatments differed according to gender. Women were far more likely to receive the most radical interventions and very heavy doses of medications. There were also more likely to die. It took just a few minutes to perform a trans-orbital lobotomy— a matter of administering local anesthesia and hammering a kind of pick, through the eye socket, a few inches into the brain.

Mental hospital populations rose dramatically through the 1950s. For women, depression—or what doctors judged as an excess libido, or even an unhappy marriage—might be enough to have one's frontal cortex severed. Following World War II, women like my mother were told to return to the domestic fold of child-rearing and homemaking once their men came home. For many, the dream of blessed domesticity never came true, and the rising incidence of divorce left hundreds of thousands of women living precarious lives.

My mother had no medical insurance. She suffered from depression and alcoholism. Her poverty meant that she was unable to seek care outside the state. Uneducated and poor, like so many other women of her generation my mother depended on men. She had lived through the Great Depression in desperate poverty, had hoped to marry a man who would be a provider. She would be a homemaker, a wife and a mother. With divorce she became an anachronism. She could not go back to get her high school degree, had no skills that promised a career, or even a modest income to support a family. My parents had always lived on the lower edge of middle class. Divorce pushed my mother over a precipice.

Mom knew about Mandeville from her sister. My aunt Cecile had spent nearly three years there before being transferred in 1957 to the State Insane Asylum, just east of Jackson, Louisiana, one of the most notorious hospitals in America. Even today, Ms. Washington tells me, everyone knows to avoid Jackson. Mom knew what might happen to her. Mandeville had a specially designated room for performing trans-orbital lobotomies. Another room was for "shock" therapies, including one in which terrified patients were brought near to death with insulin then saved by the doctors and nurses, six days a week for up to ten straight weeks. Another option: twenty or more sessions of electroconvulsive therapy (ECT), one every other day. The hospital widely prescribed psychoactive medicines such as the powerful antipsychotic chlorpromazine, which made my mother feel dumb and reduced her to the classic "Thorazine shuffle." The 1950s and early 1960s saw an explosion of medicines for the mind, though doctors and scientists still knew relatively little about the mechanisms of brain chemistry.

Between her attempt on my life and the failed suicide, my mother's world finally unraveled. I understand now that when she looked at me, her young son, she saw failure and hatred and everything that had gone wrong in her life. In the New Orleans Civil Court I discover dates that help put the pieces into some sort of alignment. She attempted to drown me around the time my father finally left, on January 5, 1963. Two years later, exactly three weeks after leaving Mandeville, at 5:20 p.m. on a Thursday evening, a court employee walked up the stairs, knocked on the door, and served my mother with divorce papers.

Dad offered to provide $200 in child support. Mom hired a lawyer, who asked the court for an extension, which the judge denied. On the eighth of March, the lawyer petitioned that my mother receive $75 in alimony, and $300 in child support, arguing that my father had kept the family "in a constant state of instability and turmoil by making repeated unnecessary changes in the family domicile." Mom returned to this very issue during our interview, her voice growing angry as she recounted all the times Dad moved the family from one rented house to another, always promising that better times lay in the next job.

The final judgment decreed a divorce and granted custody of the minor children to my mother. The judge ordered my father to pay $46 per week in child support, less than he had originally offered. This sum—$184 per month, $2,208 a year—was far below the federal government's 1965 poverty threshold. Rent and utilities left us with less than $75 a month for five people, not including my grandmother: $15 a person per month, fifty cents a day.

Nearly a year later, the court served new papers. Dad

was now asking for permanent custody. My mother, wrote the attorney on Dad's behalf:

> is not the proper person to have the care, custody and control of these children for the reason that she is unable to provide for them a proper environment for their rearing and upbringing, that she neglects them, that as a result of her actions, they are subjected to influences which should not occur in the life of my minor children.

The judge did not grant my father's wishes, though he recognized my mother's neglect. Instead, the judge "granted the permanent care, custody, and control of the minor children" to my sister Marie, by this time twenty-one years old and with two children of her own. In fact we never moved in with my sister. One child after the other headed to Mississippi to be with my father. I stayed behind in New Orleans with my mother and grandmother, entering first grade at McDonogh No. 14 in the fall of 1966. Mom began drinking soon after she got out of bed in the morning. The apartment was a disaster, fetid and crawling with roaches and mice. My mother brought men home. One man, a decade younger than she was, an alcoholic and the brother of Kinta's best friend, sometimes beat her up. A pregnancy followed, then a botched back-street abortion.

Despite all my research, I cannot see those early years much more clearly than before. I once hoped that with enough effort I would be able to open some shutter that had kept the past closed to me. I was trained in the belief that once I knew context I could attach meaning to people's actions. This is what historians do. But the past is more complicated and inscrutable than that. Context is of our own imagination, and yet it's as much a part of the histories we tell of our selves, and others, as the events we highlight. I want to be able to say that by circumstance and illness my mother was unable to care for us, unable to love. This all seems true, and at the same time not quite true enough. Something gets lost in explanation.

And yet, somehow, the simple bits of information become important. Time's stubborn specificity reminds me that while some things are gone forever, the past lives on beneath the surface of our lives, silent and powerful. There are dates, even a chronology: November 1964 to January 1965. I know now why during the holiday season the world seems drained, turns a bit more fitfully. Amid the merriment and presents and food and trees and tangles of lights I am still four years old, and my mother is in Mandeville. I am waiting for her in an Uptown house on the other side of the lake.

—⁕—

I have returned to Jacksonville a number of times over the subsequent years, but never to interview Mom again. We often sit the two of us on the back porch, just as we did that summer day. I watch the pool's reflection dance across the wooden fence and wonder at the children playing innocently in the water. We talk about the little things in our lives. Mom asks about my family, current events, who I will be voting for in the upcoming presidential election. She never asks me how the research is going. I never tell her either. I never tell her I went back to the house and my closet room and that bathroom with its tiled floor. I don't describe my trip to Mandeville, or to the courthouse with its divorce records archiving the pain of others. I suppose I would tell her if she asked, but I am glad she doesn't. I have the facts at my fingertips. What I do not have is the words to string them together.

I want to say that I understand now, understand in ways that I could not have as a child, certainly, or even just a few years ago, before I began this journey into my family's past with a simple interview, sitting on a Florida porch. Perhaps this is all the discipline of history can offer, the possibility of seeing things just a little bit differently, the unexpected angle of vision that begins emerging from our labors among the dead.

I want to say that I'm sorry for what happened, sorry for us both. That day I sat there looking into her eyes I felt a certain terror, but also an ineffable connection to someone who by all accounts was a terrible mother—not exactly love, perhaps, nor even a longing for love, but the simple tugging pain of recognition. We share a past, if not always a history. She is my mother, and I am her son.

TWO

—⁓—

ESPLANADE

THE WORLD WENT TERRIBLY WRONG FOR MY MOTHER around the moment of my birth. I was unwanted, even resented; a baby's fussing a reminder of a failed marriage, even a failed life. Mom hated breastfeeding; she wanted to have as little physical contact as possible. As a child, I could not count on her comfort when I was hungry or stubbed a toe or simply felt alone and fragile. I could not count on her even being there. I lived with the haunting presence of death.

I wonder how my mother's crises, the seemingly endless injuries and the disavowals that lay within every bottle, the very breakdown of language itself, at times overwhelmed my developing brain's capacities to make sense of things. A torrent of stress hormones consolidated Mom's attempted suicide so that a fleeting childhood moment remains forever present. In the firmament of electrical signals, transmitters, proteins, chemicals, and neurons branching and withering in bewilderingly complex tangles, my brain developed an ineffable difficulty in processing experience and forming memories that distinguish the past from the here and now. Perhaps as a child I fell into dissociative states as a primitive form of protection. Perhaps also I began honing a kind of perspective, a writer's remove.

The origin of forgetting resides not only in my cerebral folds. The making and remaking of collective memory came to an end. Mom stopped telling stories. The simple rituals of family life that are so important to developing a sense of our selves and others disappeared—the meals shared, the clicking Kodak Instamatic recording celebrations and holidays, the yarns tying lives together that allow children to tell their stories and, one day, their parents' histories. An existential muteness occupied the space where once there had been the motions of family and the making of memory. Even the photographs that sat upon a table memorializing a past, evidence of a seventeen-year-old marriage, now stood alone in their frames unremarked upon, or were simply thrown out with the rubbish, a life annulled.

Which is why I've been collecting materials, sifting through records, phoning receptionists asking for forms, wanting to know when I might expect this death certificate, that medical record. I've assembled a chronology of my childhood, though for the life of me I can't commit the dates to memory, so I keep copies in my office and in my briefcase. One chronology leads to another, a mechanical attempt to find out what people did, to create some sort of conversation with the dead. There are photographs I've taken and collected, notes I've scribbled down, interviews recorded, a timeline I am trying to fill in. And a list of questions, many of inexplicable relevance: about the color of a rosary, the smell of Johnson & Johnson's talcum powder in the summer, the resolute presence of my grandmother, Mary Cecile Samuela Salvant.

I walk around New Orleans in the months after Hurricane Katrina wondering, hoping, if I might turn history into memory. I stop before cement steps leading to a hummock of weeds three or four feet high, then before an open briefcase filled with records that have been misshapen by the great storm, rendered brittle and unreadable, a black mold creeping over the words. Large parts of the city seem like a ghost town, but then New Orleans has always been occupied by the living dead. Standing outside the apartment, images form as if they resided just past the screen door, indistinct but definite traces of some other time. I am left with the disquieting feeling that I will discover that behind everything lies something else that is both present and yet inaccessible, immensely powerful by its very absence. I record these phantasms, test them against my notes, expand the chronologies, try keeping an appointment I have with the past.

I imagine my grandmother is still there by the window looking over me. In the swirling silences of childhood, Grandmother appears as clear as winter's light, forever brushing her hair, whispering prayers in a nightgown of gossamer-thin cotton. A hand moves up and down arms weakened by time and history, then around the back of her neck and across her bosom as she tends to her sinewy remains.

The talcum powder keeps the dampness away, she says, makes you feel cool and fresh. A brush descends through long hair, which she then braids and fastens with a red rubber band from the *Times-Picayune*. The ponytail goes straight down the vinyl back of the wheelchair. A crimson and gold rosary falls across her fingers. Beneath the sweet, slightly floral talcum powder, there is the sour-milk smell of the aged.

Grandmother lost a leg from a small wound she had left unattended, believing that long hot baths with Epsom salts might exorcize the pus from her body and allow her to avoid paying for a doctor she could not afford. She finally went to the hospital, but it was too late. The surgeon took the leg off right at the knee, stretching a flap of skin over a wound that would come to be held together by three scars at neat right angles.

She moved into the apartment in July 1965, age eighty-three, six months after my mother returned from Mandeville and just a month after the divorce judgment, taking the front room, my parents' room, though I have no memory of them together nor of the time when I was three and my mother slept there alone.

Against one wall Grandmother created a small altar with hollow porcelain statues of Jesus and Mary. Father Canon arrived from Saint Stephen's on Sunday afternoons when Grandmother turned her room into a chapel. My sisters remember meeting Father Canon with candles and guiding him into Grandmother's room. From the corner of my eye I can see her bending down on one knee to receive the Communion. I was five years old, soon to begin kindergarten. The monthly Social Security check helped pay the rent and groceries. Grandmother taught me French and "the three R's." She soon had me memorizing multiplication and division tables, one through twelve, repeating them as one would practice musical scales.

Grandmother kept a list of important dates: births, marriages, deaths, and announcements, carefully excised from the newspaper and taped into a notebook, the "Family Record." The past offered some solidity, some staunchness in a life of poverty and insecurity and dissolution. She was the family archive, for decades offering stories about a distant past (though I cannot recall any of them), as if recounting good fortune presaged its return as surely as one caressed a magic lantern. Once the family had been rich from the cotton business, wealthy enough to have a slave serve dinner in their house on Esplanade, which was one of the city's great avenues, where the Creole elite lived at the height of New Orleans's prosperity. It was a grand house filled with polished European furniture and a door of cut crystal. Through the man-high windows, one could hear the great steam paddle boats trumpeting their turn round the river bend.

—⁓—

"Dontcha remember when Michelle stayed with us," Sabrina asks me, "our cousin?"

"No, I just remember Michelle from your wedding."

"Seriously? I remember Michelle screaming 'They killed her! They killed my mother!'"

"No, I don't remember anything about this at all. Nothin'. Really, Sabrina, I think I remember Louis or Teddy [Michelle's older brothers] carrying Grandmother down the stairs once. And I know we lived off her Social Security check. But I don't remember anything about Michelle living with us. I only just found out that Cecile had died that summer." That was 1965, just months after our mother's attempt on her life. "The only thing I remember is Hurricane Betsy and how the streets flooded and the chinaberry tree fell right across the road. I don't remember Michelle living with us at all. Are you sure?"

Sabrina seems perplexed that she has so clear a memory drawn against my blankness.

"Well, Michelle was convinced the hospital had killed her . . . Like they had done something to her brain. I remember her saying, 'They killed her! They killed my mother!' She stayed for a while, like a month or so."

I explain that I am trying to find out what happened to Cecile, my mother's sister. I want to know as much as I can of our mother's life, her family, what happened. Many records have been destroyed. There are confidentiality issues too. Mom said she would help. She's the closest living relative. Maybe I'll find out something. Then I describe what I do know about Louisiana's mental health system in the 1950s and 1960s—why women were treated so differently, the kinds of procedures doctors performed on patients, the sorts of medications they administered: insulin therapy, ECT, frontal lobotomies, Thorazine, Stelazine, Mellaril, etc., until Sabrina stares at me and I half-expect her to ask "Why do you want to know all this?" I am not always sure myself.

Certain facts come easily, as in most historical research. It is not difficult to track down a birth or death, the events dutiful clerks register in the state's ledgers. The Internet has revolutionized genealogical research. There are multiple databases—some free, others for a fee—that make it extremely easy to create family trees. I simply type in a first and last name and out pours a wealth of genealogical information. Federal census data also has been digitized, making it simple to identify residences and household composition. Many newspapers are available online, allowing one to whizz through them by using keyword searches. Today, everyone with a computer can become their family's genealogist.

It's all the in-between things that are difficult to discern—the innumerable motions of lives that so often turn into time's infinite ashes. Historians delight in discovering materials that were not meant for public view; by these sources we believe we can begin listening in on people's interior worlds, in short, on their consciousness. I have no diaries or letters to work from, no treasure trove by which I can begin understanding my family's history. I have to cobble together an archive from so many bits and pieces: a police record, a divorce decree, and in Cecile's case a thick medical file recording her insanity. I have been trained as a social historian, so I am used to working with fragments. It's labor-intensive research. You have to be willing to scour every kind of source, and you have to be patient. A week of research might yield just a few notecards of information. You take what you can find and move on, though it's also important to reflect on what historical forces have produced the evidentiary record.

Cecile was born in Texarkana in early 1915, during the Great War. Zeno kept the family just out of destitution by traveling though the Ozarks and northern Texas and Louisiana, hawking texts for the Children's Educational Books Company. He stayed away weeks at a time, always returning home slightly forlorn and never with enough money to settle down for long. They returned to

Louisiana—first Baton Rouge, then back to New Orleans. Cecile grew up in various shotgun houses Uptown, mostly in working-class areas of Carrollton, never very far from the sound of trains atop the levee and ships plying the great river. Grandmother thought of her as an easy child with an "even" temperament. Teachers complimented her good work, fine memory, and most of all her yearning.

Grandmother raised her children on stories of the great house and how life had been when there was money and good fortune. One could live by stories, if they were told often enough. Stories sparked into fantasies of family walks down the Esplanade past homes with cut-crystal doors that led into large square rooms filled with mahogany antiques, a rainbow's light dancing across the floors. Servants stood in thick, pressed cotton uniforms serving meals on tables of burnished wood covered in white damask. At winter's end gowns of red taffeta twirled at the Proteus Ball, and flames danced from the Bananas Foster at Galatoire's.

Wealth would one day be hers—of this Cecile was certain. She would be a proper woman of means. While everyone else drank café au lait, Cecile sipped tea from the finest china she could find. It was a small but important way of reminding herself of the English blood murmuring through her veins, from distinguished and rich descendants on her father's side.

Cecile spent her teenage years and early adulthood in the Great Depression. She thought of becoming a nun. For a short time she entered a convent but found it all too depressing and returned home, where she spent long hours bent over the radio and going to the Ashton Cinema whenever there was enough spare change for a ticket.

Cecile graduated from high school during the height of the Depression. She took various jobs—governess, saleslady, beautician, seamstress—until 1937, when she quit and married a Cajun man from Lafayette. Old Man Louis—that's what my siblings and I called him—enjoyed weekend evenings picking through big succulent boiled crabs piled on top of a *Times-Picayune* spread across a Formica table. In the decades after World War II, when many whites fled the inner city, Louis made a good enough life building homes out in Gentilly and in the other New Orleans suburbs, barking orders in a barely intelligible accent with a wet, chewed-up cigar hanging from his mouth.

A child arrived soon after their marriage, then a year later another, both sons. Cecile had five children in all; the last, Jeanette, was born in 1952.The couple rented a non-descript shotgun on Cambronne Street, a few blocks from the levee and not too far from her mother. In the 1930s and 1940s this area housed what elites impolitely called the city's white trash, people who worked by the sweat of their brow or moved from one job to another within the city's expansive informal economy. Today the house has been lovingly renovated as part of the post–Hurricane Katrina return to Uptown by middle-class whites looking for areas that won't flood in the next great storm.

The couple argued. She deserved better: a large house in a posh neighborhood where she could bring good company, hold elaborate dinner parties, drink Ceylon tea from bone china. Descended from American aristocracy and the Creole elite, Cecile felt entitled to more than being married to a Cajun contractor. She knew the history of her mother's family: Joseph the wealthy cotton broker, the great house on Esplanade. Her great-grandfather's last name was Adams. "We're related to one of America's founding fathers, and before that to someone who had come to this country on the Mayflower," Zeno had told his daughter. Somewhere in America there were the "right relatives."

For months, Cecile picked up the phone and began dialing anonymous people across the country, trying to discover her rich relatives living in stately homes and settled in their economic fortunes. Louis had the phone removed.

There are so many silences within the written archive, particularly when the records deal with the most intimate aspects of people's lives: love and jealousy, sex, a woman's dreams, madness. It is difficult to discover what really happened, how and why my aunt went crazy. Some facts are easy to come by. In 1943, just six years into her marriage, the police arrested Cecile for creating a ruckus at her parents' house. She also suspected that Louis was a philanderer. Cecile started following her husband around, yelling at the women Louis was working for to "leave my husband alone."

She neglected the children. Their names begin appearing in the police records between 1948 and 1952: an eleven-year-old daughter reported missing, notations by officers of arriving at the house because Cecile was disturbing the peace. The police took Cecile to the hospital. The children went to the precinct headquarters.

By October 1951, Cecile was spending hours and hours in her room before the mirror, brushing her hair and laughing, as if she were preening herself for a lavish party. Cecile would go downtown and buy the best clothes at the most expensive shops, then go home and within a few days give them away.

Cecile would also take the Carrollton trolley to City Park, and then a bus along Esplanade, until she found the old house. For hours she would sit on the front porch, content that she had returned home. "Get out! Get out!" Cecile screamed when the owners returned. "This is my house. You have no right." The police came and drove her down to the Seventh Precinct, and then to the mental ward.

For the next thirteen years my aunt was in and out of mental hospitals. Mostly in. During Cecile's first hospitalization, doctors administered twelve ECTs, followed by a course of insulin shock therapy that involved thirty-six episodes and thirteen comas. She also received the diagnosis of paranoid schizophrenia from which she would never escape.

Louis disavowed his wife. By 1954 they had divorced, their youngest still in diapers. For some time the children lived with my grandmother and Zeno at 7901 Cohn Street. It was a long walk from their home on Cambronne, but near enough to their father to help soften the fact that the family had broken apart.

"My place is home caring for my family," Cecile pleaded with the doctors. She had to "return at once to the city and assume my responsibility for the three younger children."

Each time Cecile returned to New Orleans the hallucinations returned as well, along with arguments with Louis and her parents. Jesus sat in some Uptown shotgun, held against his will. General MacArthur governed from the White House. Everyone plotted against her.

Cecile spent time at DePaul and Charity Hospitals before

making the trip across the lake to the mental institution in Mandeville. She received further courses of ECT and insulin shock therapy. Thorazine, Stelazine, Mellaril, and the powerful tranquilizer Tolnate entered the arsenal of treatment. Cecile escaped more than once, each time trying to make it back to New Orleans and to her children.

In 1957 the family agreed to have Cecile committed to the state hospital in Jackson. Built in 1848 during a great wave of American incarceration, the Louisiana Insane Asylum had among the worst conditions of any hospital in the country. Patients were left in their own excrement, wandered naked through the corridors, suffered terrible malnutrition and neglect. Doctors routinely administered electric shock and insulin shock therapies. Frontal lobotomies were widely performed on all sorts of patients, especially women, until the emergence during the 1950s of powerful tranquilizers and antipsychotics.

Mrs. Gremillion, a Jackson hospital doctor records, brought with her "a disorganized letter which she sites [sic] as evidence of her concern with the welfare of her small children." She also brought newspaper clippings of police and searchers standing next to the mutilated body of a six-year-old girl.

"Why are you here?"

"You tell me," Cecile replied, explaining to the doctor that she was well known in polite society, descended from the very best stock of Boston and New Orleans.

She had, as it were, "a superior air about her."

Cecile kept trying to escape, mostly unsuccessfully. She was discharged in 1958, but returned to Jackson in 1959 and moved to a locked ward. At one point she was on the highest doses of antipsychotics. It became difficult to wake her up in the morning. Lethargic, wasting away, she appeared to one doctor as a "zombie," though she still insisted on her "family responsibilities." Cecile began having difficulty walking, increasingly falling down. She accidently injured a finger in an electric fan. The doctors worked her up for another course of ECT in March of 1964: "She would need at least twenty if not more." Through June, July, and into August they wheeled Cecile into the room, blasting her brain with electricity. Her condition deteriorated further. At the end of the year Cecile's daughter filed court papers to declare her mother legally incompetent.

A routine medical exam in 1963 had revealed something in Cecile's left lung. The doctor suspected a minor infection, but her condition steadily declined. Late in the evening on August 22, 1965, soon after being transferred to Charity Hospital in New Orleans to determine why her health continued failing, Cecile died. Undiagnosed cervical cancer had spread to her lungs, brain, and adrenal glands. She was fifty years old. A few miles away the judge had just ruled on my parents' divorce.

Two weeks later, Hurricane Betsy hit New Orleans in one of the worst storms to ever hit the city. The tempest stretched nearly six hundred miles across and had a forty-mile eye and 120-mile-an-hour winds. Betsy hit the city at night, tearing off roofs, knocking down trees, breaching some of the levees. The flood waters consumed the crypt that housed Cecile's remains.

Louis and his children filed a wrongful death suit in July 1966. They believed the doctors had killed Cecile, alleging that the mental hospital had failed to provide the most basic medical care that would have identified the cancer long before it had spread through her body. The First Circuit Court of Appeals ruled on the matter in the summer of 1972, affirming the lower court's ruling and dismissing the case.

—⁂—

I wanted to know more about Cecile's life, because I thought that through my aunt I might better understand the women in my family—my beloved grandmother and especially my mother at the time during the early 1960s when her own world fell apart and she ended up in Mandeville. The research leads me in unexpected directions, and makes me feel as if I am compelled to follow some labyrinthine trail into the past. There are newspapers to read, obscure documents to collect, even Google Earth to look at. With two clicks of the mouse, I am staring at one of the shotguns my grandmother rented before she lost her leg and moved in with us on Chestnut Street.

Historical research is often this way. We follow leads as we sniff around the past trying to make connections, triangulating bits of evidence to reveal possibilities that remain silent within a single document. Historians, particularly social historians, hope to tell stories of peoples otherwise condemned to silence, scouring the past to bring the dead out "from the enormous condescension of posterity," in E. P. Thompson's memorable words. We might bequeath to the future some sort of wisdom from the gritty details of those who struggled, suffered, and so often prematurely died. This seems important. And insufficient. I find myself not just wanting to know what happened. There is also a yearning to converse with the departed, to convene all these ghosts for some kind of colloquy, a communion of tenses.

I dreamed of finding an empathic ghost, someone who cared and most of all listened, a kind of witness to the lives of others, who would be able to give me something knowable about my family's interior lives. In one record a social worker expresses concern for the stress to Cecile's children. I had hoped to find more like this, but in fact, throughout Cecile's decade-long struggle with insanity, no one listened to her story. They asked questions, certainly, and Cecile answered, sometimes in detail and with enthusiasm, according to the doctors. She *wanted* someone to know her story, including all the demons, wild ideas, and concerns that swirled about her: the Uptown Jesus, MacArthur in the White House, Eddie Fisher's betrayal of Debbie Reynolds. Her children were in danger. They needed their mother. Louis had been unfaithful. People owed her money. She had the good arches of a proper lady. Once there had been a house on Esplanade.

The doctors would be wasting their time listening to a woman's mad ravings. After World War II, schizophrenia became a quintessentially female malady, despite evidence to the contrary. Women were no longer hysterics. Now the problem was that their minds were broken, quite literally—*schizo* ("split") and *phrenic* ("mind"). In an age when Freud was becoming a household name, the medical establishment believed that schizophrenics, unlike neurotics, had no need for talk therapy. It wouldn't help, so there was no reason to listen. Madness and narrative were distant relatives, and not on speaking terms. It didn't matter to the doctors, the lawyers, not even to the family, what Cecile had to say.

Schizophrenia demanded the full battery of the psychiatric profession, from powerful sedatives and antipsychotics to frontal lobotomies. That first diagnosis of schizophrenia stuck. Each doctor's summary and notes borrowed from the last in a long chain of records from the specialists who lobbed one medicine after another at her, to those who strapped her to a gurney and brought her to death and back, to those who zapped her with enough electricity to power a small town.

I decided to listen to what my aunt tried to say through the paralyzing tonic of psychopharmacology, wondering if what I found might take me out of the hospital to other materials, different sources. It's rare for the historian to rely on a single source, and it's usually a bad idea. Records come with their own silences and mysteries. People often willfully suppress something important, producing within the document a powerful absence. Like memory itself, a diary or a police report or my aunt's medical records exist by rules of selection and consolidation. Historians train themselves to identify what's missing—and why—as much as they interpret what they've discovered.

The records that exist like a penumbra around my aunt's life begin revealing darker secrets the family kept to themselves. A few weeks after Cecile first entered the hospital and the children ended up at the police precinct, Old Man Louis and Cecile's brother, Henry Mullan, sat in a corner bar on Dante Street just around the block from the shotgun on Cambronne. They walked out of the bar mid-afternoon to Louis's car. The two of them got in, closed the doors. Henry reached into his coat and pulled out a loaded .25-caliber Colt automatic pistol. They drove around for a while, the gun trained on Louis. They likely argued, perhaps about Cecile, until the car needed gas. While Louis stood at the pump, Henry got out and ran away. The police arrested him on Jefferson Highway, booking him for attempted murder and for concealing a loaded weapon.

The psychiatric records tell me that Grandmother took in Cecile's children, helped feed and clothe them with the little money she and Zeno had. When Cecile was discharged, or when she returned to New Orleans on a short leave or during one of her escapes, before the police car arrived and drove her back to the madhouse, she lived with her parents. Their relationship turned fractious. Mother and daughter had terrible, violent arguments.

I cannot discern the content of their fights from the police and psychiatric records. Were they about Cecile's children, her marriage, the character of my aunt's relationship with her parents, or something else? What were the voices telling her? It is impossible to say. There exists but the faintest hint that Cecile had grown frustrated with her parents' poverty. "My parents aren't white," Cecile once told a doctor. "They're colored . . . My mother works like a slave."

Or was there some wound of neglect? When their daughter needed them most, my grandparents were not there. For seven months in 1954, neither of them bothered to inquire into their child's condition. When my grandparents finally made the trip to Mandeville, the doctor was amazed they had not once looked into the plight of their daughter. They struck him as "obviously disturbed individuals." My grandfather arrived "very unkempt and unshaven," and "dressed in some sort of uniform." Grandmother wore an eccentric outfit "suggesting the year 1910." During

Cecile's years languishing in the squalor of Jackson, they never once visited their daughter.

This discovery troubles me. I want to say that I could count on my grandmother, that she was the only sensible person around, that she worried about me and that she gave me an education, somehow instilling the importance of learning and hard work. I believe she saved my life. "Grandmother," I want to say, "I can see you still, sitting in the wheelchair, the rosary beads slipping through your hands as you watch over me. You had been as constant as the morning." My siblings tell me that our grandmother watched me from her bedroom window while I played in the ditch near the house. But now I wonder if I haven't taken their relic and turned it into a memory and a story of childhood, none of which may be true. I have conjured an image of my grandmother from the cultural archetype of the benevolent, wise elder shielding her young charge from the foolish and often dangerous world of adults.

Other puzzles and fantasies begin emerging. Early in adulthood Cecile had seemed a bit different. The second decade of her marriage had been especially difficult, filled as it was with children and little money, jealousy, and fantasies that had begun ruling her life. She began losing her mind definitively around 1951, when she entered DePaul Hospital and the doctors filled her with drugs and shocked her brain. What is unnoted in any of the documents is the fact that a few months before she became a patient Cecile had become pregnant. She spent months alone in the hospital with a child growing inside her and with no one particularly interested in what she had to say.

Following her release, her youngest child an infant, Cecile occasionally ventured into the city—not to Esplanade and the old house, but to a Woolworth's store. There my aunt made a spectacle of herself, screaming and accusing the store manager of being the father of her baby.

Cecile repeated this accusation until the medicines made it impossible for her to know who she even was. No one ever believed Cecile, or if they did they kept it to themselves. No doctor or nurse or lawyer believed Cecile either. Nor even her mother and father. It didn't matter what she said. She was crazy, stark, raving mad, even though my aunt gave a precise name to the man she accused of being her child's father: Arthur Mullan.

"Have you ever heard of an Arthur Mullan?"

Mom's never heard of him. No one in my family seems to know anything about him. I start sifting through my records, coming across a genealogy Mom had begun many years ago, and there he is, my grandfather's youngest brother. Arthur Mullan was Cecile's, and Mom's, uncle.

I call Mom again, now suspecting that something loathsome happened to my aunt. Had she been molested, even raped? Mom begins to remember, acknowledging his existence but adding very little. Men move on, she tells me. By marriage families lose sons but gain daughters. Men are always disappearing, wandering into someone else's life, wandering away. I find her answer unsatisfactory, even irksome, though I realize I myself have drifted away. Independence has always required a certain distance, even a measure of disavowal.

Searching for Arthur Mullan leads to census records, police reports, reels of microfilm whirling before my eyes, even a plane trip to Salt Lake City and the Mormon's Family

History Library with its vast repository of genealogical records. There's not much to go on. Crucial documents seem incomplete, even contradictory. It doesn't help that he switched around his first and second names. I learn he qualified as a pharmacist in 1901 following two years of college; in 1909 he married the thirty-one-year-old Cora Peterson; and in 1915 their first child died while the couple was living in New Orleans. Arthur registered for the draft at the end of World War I before heading west, first to Arizona and finally settling down in central Los Angeles, where the couple rented various houses and Arthur plied his trade as a druggist. He died at the end of 1942; depending on the record he was in his fifties or sixties. Cecile would have known Arthur only as a young child of six or eight years old. Her uncle was long dead by the time she became pregnant with her last child.

The research leaves me feeling bereft, and a little troubled. It's not that I've wasted too much time on a wild goose chase. Or that a hypothesis has turned out to be incorrect. Developing theories of causation is at the center of much historical work, and theories are meant to be tested. Rather, I've committed a common error that befalls historians: the belief that we can discover an origin that explains everything else. I wanted to be able to say that incest led to my aunt's insanity, as if I could draw a line connecting two dots separated in time. I would engage in the recuperative project of history, figure out what happened, and understand the tumult of her heartbroken life. I was more than willing to jump to conclusions. All I needed to do was uncover an unsavory family secret— a single traumatic moment—and all else would be explained. The rest of Cecile's life would fall neatly into place, however tragically. Now it seems impossible to understand my aunt's life. In some basic, resolute way, her past, perhaps every past, continuously slips away.

This research also leaves me feeling dirty, not only from prying into the lives of others but by association—too close to a chasm of tragedies from which I want to escape but seem instead to be falling into. Cecile was undoubtedly insane: she heard voices and spoke to them; she thought she "might be her own twin"; she combined "bizarre sexual" and religious ideas; she was filled with "grandiose delusions," "preoccupied with self-improvement, education and seems to have a great deal of underlying insecurity." Her dead siblings were living in the house with Jesus, who had assumed the name of "Dr. King." At times she disowned one of her children, at other times insisted she had six children, not five. Arthur Mullan becomes her brother, a brother murdered by her husband in a fight over a $60,000 house he had constructed for his secretary and lover. Everyone in my family seems a little crazy, though: people pulling guns on each other, jumping off bridges, dressing up, lying down on the bathroom floor and turning on the gas. The madness never ends.

I find myself wondering if my aunt's delusions have some deeper location in the stories and longings of her mother, my beloved grandmother—stories that led Cecile to a house on Esplanade. Were all the performances—the outbursts, the clothes, the phone calls and the trips downtown—actually embodying an unspeakable past, a past not known but somehow remembered? Did madness steal into a space that language abandoned? I wonder what has been lost, even silenced, in the making and remaking of family memory.

I work backward from what appears settled to ever more distant and murkier times. Published census records make it relatively easy to identify where Grandmother lived, household size, occupation, and whether or not her family rented or owned their residence. Soon I am poring over address books, registers of birth, marriage, and death, wills, church records, phone calls and emails to genealogists, and scholarly tomes on the early history of Louisiana. I keep expecting to find the house on Esplanade, as well as information that will explain how the family fell from the Creole bourgeoisie into the ranks of the urban poor white class. There are a few precious family documents Grandmother kept that stretch back into the nineteenth century, which include a list of births and deaths and where people lived, and a miscellany of writings in English and French: poems, stories, a few French ditties. The trick is to place these materials into a kind of conversation. The easy part is identifying instances where one source confirms another. The more difficult challenge is understanding the relationship of the written archive to the storehouse of family memory.

Christened Cecile Samuela Salvant, Grandmother was one of ten children, a few of whom did pretty well. One sister moved out to California with her husband, an Irishman named Sweeny. During the Depression she sent bags of clothes back home to New Orleans until her husband ended up in federal prison convicted of fraud. Another sister, the beautiful Mae, opened up a typing school for secretaries. Mae lived not too far from Esplanade, in a double shotgun with a yard filled with peach and fig trees.

Cecile Samuela attended the Louisiana Normal School for Teachers, working through her twenties until she met

Zeno Mullan, a ne'er-do-well son of a Scots-Irishman. The Mullans were all working-class New Orleanians: peddlers, grocers, shop employees, and common laborers, some of whom at least pretended they were descended from American aristocracy all the way to the Mayflower.

My mother's parents married and had their first child, a daughter, at the beginning of 1911; Grandmother was then twenty-nine. Louisiana law forced women to give up teaching once they married, so she lost whatever financial independence she once had enjoyed. The couple lived week to week and fought often. Zeno wandered from one job to the next. Sometimes he simply disappeared. At the height of the Depression he abandoned the family for a few months, hoboing trains to the Chicago World's Fair. He ended up working as a night watchman at Charity Hospital until his death in the spring of 1957.

Children offered little solace to this broken marriage. Grandmother lost one child in infancy, a son barely six months old. Her oldest and beloved daughter Rowena died at eighteen from a burst appendix. In June of 1946, her son Russell jumped off the Huey P. Long Bridge. He was twenty-nine, married, father to two daughters, and had just returned from the war in Europe. For a while Russell, too, had spent time in a mental hospital. This left her with her namesake Cecile, Henry, and my mother, Yvonne.

Of six children, only two survived Grandmother: my mother and her brother, both alcoholics. Henry had gone to war divorced and with a fondness for drink. He returned to New Orleans and remarried, took up odd jobs, lived in various apartments in the city, and slowly drank himself to death. Henry ended up escaping the prison where he'd been sent for pulling the Colt on Louis in 1952, and in later years he managed to eke out an existence Uptown, in the end running one of the city's many tiny neighborhood corner stores, where people bought dry goods, cool drinks, beer, and cigarettes. When I met him in the late 1970s, Henry was hallucinating wildly and holding elaborate conversations with a Chihuahua that he kept in his sweater in the middle of the New Orleans summer. A robber had blown half his head off. Most of his lower jaw and one side of his face had collapsed in on itself, so it was impossible to understand much of anything he said. He looked like a Lucian Freud painting, with the asymmetries of his face and slackened folds of flesh. He died a few years later of alcoholism.

Grandmother's personal history feels as if it is easy to discern. I can conjure within my mind a sense of the Depression's hard years and the decades of living month to month, perhaps because that world, bequeathed to my mother, for some time became my own. I wanted to know though about what had come before, the splendor contained in the stories of a more distant past that had led Cecile to someone's porch on Esplanade.

"Well, son, I remember my mother telling me about Joseph the cotton broker," my mother tells me. "He was rich! Real New Orleans elite. Aristocrats. They had one of those great big houses, and a slave who served them dinner."

"I remember Grandmother telling me she lived in a big house," a sister informs me, her eyes glistening. Stories passed from one generation to the next declare the Salvants to have been a family of good, pure, French stock—solid New Orleans Creoles, as my grandmother had told her children and everyone else who would listen. Her father, Joseph

Moliere Salvant Jr., has been a comforting apparition to those generations who have lived by their wits but remain forever certain that good fortune lies both in the past and just around the corner. Mom talks of Joseph as if he were sitting by her bed dressed in a fine suit and hat, describing his life as a cotton broker. She descends into a time when the labor of slaves created a New Orleans grandeur of fancy mansions and servants tending to the family's needs, resplendent masked balls, banquets that went into the morning hours, and cases of precious Bordeaux and Burgundy wines for the fine cellars of restaurants like Antoine's.

Joseph Jr. was one of eight children born to Marie Josephine Veillon and Joseph Moliere Salvant Sr., the namesake and treasured oldest. I discover that his childhood in the 1840s lay not in New Orleans but on a small ribbon of land in the swamps of Plaquemines (Atakapa Indian for "persimmon") Parish along the west bank of the Mississippi River, twenty or so miles south of the city.

Going further back in time I learn about Jean Salvant II, the father of Joseph Moliere Salvant Sr., who was born in Plaquemines in 1785. He married Rosalia Daubard in 1803, whose father had migrated from Bourgogne and, in the New World, taken Rosa Charles as his wife, a woman of likely mixed-race descent. Jean passed away by 1827 when Joseph Moliere Sr. was twelve. Little is known of the pioneer Jean Salvant I. He arrived in Louisiana as a soldier during the middle years of the eighteenth century, probably from central or eastern France. In 1770, he owned just over three acres of land in Plaquemines granted to him in return for his military service, along with four hogs and a rifle. He married Maria Luisa Lambre. In 1796, one of their sons sought a special dispensation from the church in order to marry his cousin, a common practice in early colonial Louisiana.

Plaquemines was a poor, rough, backwater area along the river that stretched down to the French fort and settlement at La Balise near the Gulf of Mexico. Early colonial farmers cultivated the black soils along the river bank, planting corn and vegetables that they sold whenever they could in New Orleans. They built small houses of swamp cypress and yellow pine with broad roofs to protect them in the hot summers. They hunted and fished, raised a few pigs and cattle, and in the cooler months stood in pirogues raking the legendary Plaquemines oysters that hid in the muddy, brackish waters. Experienced river men helped ships navigate the Mississippi's treacherous and ever-moving sandbars. Others turned to pirating and robbed boats bringing goods up to New Orleans. Most lived and died poor.

For the early pioneers Plaquemines was a foreboding, primordial place, more Africa than France. Fevers from the miasma drifting out of the swamps and sea marshes took many colonists to an early grave, especially the young. The "infinity of mosquitoes" turned the night air into an incessant drumbeat. Settlers had to contend with alligators, water moccasins, black bears, and outlaw pirates, who hid in the back waters waiting for boats moving up and down from New Orleans.

Most of all they worried about the weather. In the summer thick courses of raven clouds formed along the southern horizon. In the great hurricanes the ocean rose into the sky. The sea would fall into a tumult. First the light would turn a buttery yellow, and the wind would circle around the top of the trees as if it had lost its way. Then clouds would rush over the forest low and heavy, and the rains would turn roofs into beating drums. Pine trees whistled shrilly until the world became a roar, with trees uprooted and the sea marching inland to sweep the innocent away.

For a short while, especially during the Spanish period (1762 to 1802), the acrid smell of indigo leaves boiling in vats drifted across Plaquemines, the plant's radiant blue dye destined for European finery. Indigo brought slavery. For a few colonists it brought an early taste of wealth. The parish became a gumbo of people: colonists, virtually all Roman Catholics from France, marginal men now free from feudal restraint, a few from Spain, Cuba, and Mexico; slaves brought directly from Senegambia and Congo or from the Caribbean; free blacks; and the maroons who joined the few remaining Chitimacha Indians deep in the swamplands.

Southern Louisiana had a uniquely high degree of miscegenation, particularly in the second half of the eighteenth century. Louisiana has one of the most complicated racial histories in the United States, more akin to Cuba than to the other Southern states. Plaçage they called it, left-handed marriages, *mariages de la main gauche*, involving European settlers, Creoles, slaves, and Indians. Miscegenation produced mulattos, quadroons, and octoroons in the colony's finely graded racial system that was remarkably different than the rest of the American South. A Daubard son married a woman who descended from a "negress" slave born in Africa at the beginning of the eighteenth century; one of their children, St. Luc, became a mulatto sugar farmer and owner of some sixteen slaves in 1850.

The indigo economy collapsed within decades, done in by crop diseases and competition from India. Rice, introduced by African slaves from Senegambia in the early decades of the eighteenth century, proved more dependable and was in high demand throughout the Atlantic world. The Daubards became early rice planters, the sweat of their slaves building levees to create paddies and protect the crops from salty waters. It was sugar, however, that utterly transformed Plaquemines Parish. Sugar took off spectacularly at the turn of the century, following the slave revolution in Saint-Domingue that destroyed what had been France's Caribbean gem and sent thousands of slave owners and their slaves to the United States. Louisiana would become a leading producer of sugar, the largest in the country, creating fabulously wealthy plantations up and down the Mississippi River. The average plantation was worth more than three million in today's dollars and the larger ones were far wealthier, producing thousands of hogsheads of sugar and exploiting hundreds of slaves. Elsewhere Louisiana colonists had adopted cotton, a far more reliable crop than sugar, though less spectacularly profitable. Nonetheless, during the sugar boom many cotton plantation owners rolled the dice on the cane.

The Plaquemines population grew steadily in the early 1800s in response to the sugar boom. By 1820 the population had risen to 661 whites, 1,566 slaves, and 151 free blacks; the slave population would increase steadily in the following years. Sugar required huge investments in slaves and equipment. Hurricanes could destroy men's entire fortunes. (Ten major hurricanes hit Louisiana between 1780 and 1830, the crucial years of sugar cane expansion.) Animals ate away at the crops. Fields required constant weeding. Most of all, sugar brutalized: lives cut short by over-work, bodies pulled apart by machinery and the whip, kin torn from one another, the indifference born of lucre. Slaves resisted, protesting the brutal work and harsh discipline, often absconding at crucial moments, though in the nine-teenth century the swamps no longer offered much safety. At the beginning of Carnival season in 1811, while planters ate lavish meals, drank, gambled, and danced, upward of 500 slaves deserted sugar plantations near New Orleans in the largest slave revolt in American history. Led by Kwaku and Kwamina, two Asante kingdom warriors, the slaves marched on the city dressed in military uniforms carrying guns and cane knives, the ideas of the French Revolution and Saint-Domingue's slave revolt in their heads: *egalité*, *fraternité*, most of all *liberté*. A planter militia defeated the rebels five days later. The heads of Kwaku and Kwamina ended up on poles, along with a hundred others.

Few pioneer families had the capital to become large sugar planters. Wealthy Anglo-American merchants from New Orleans and men from across the South descended on Plaquemines, buying up most of the good land, their slaves planting the new ribbon and Otahiti cane that would make Louisiana famous, and building stately homes at the end of tree-lined roads. The fineries of success lay just outside the reach of men like Jean Salvant II. Jean became a sugar maker on a plantation near Jesuit Bend, where priests first intro-duced sugar to Louisiana. The job offered a way of escaping his father's poverty. Sugar making was an important, tech-nical job. Experienced sugar makers could make as much as $1,500 a year, an impressive salary at the time. The work in the sugar house went on twenty-four hours a day, seven days a week, for as long as three months. It was brutal, dangerous work: limbs could be caught in the mill, or men scalded by the boiling syrup. Tempers flared. Overseers carried guns and whips to enforce discipline. Jean slept in the sugar house to oversee the labor. Slaves pressed the cane and boiled the juice in a series of four to six wrought-iron kettles. Lime forced impurities to the top and was crucial to granulation. At the perfect moment, when the juice was sufficiently concentrated, Jean would order a "strike." Striking too soon affected crystallization. Striking too late altered the sugar's taste and coloration. Slaves then ladled the juice into cooling vats, where it turned into various sugars and molasses.

Jean's work on the plantation brought a modicum of improved fortune. His sons learned the world of sugar, and each did better than his father. They began purchasing slaves, though none entered the great planter class with its columned mansions and double-storied balconies. By mid-century the Salvant men in the parish owned a total of twenty-two slaves, including four mulattos, a number that was still half of what was needed to run a viable sugar farm. Jean Salvant III owned eleven slaves in 1850; he became the most prosperous Salvant, with real estate valued at $60,000 two decades later, a small fortune equivalent to more than $750,000 today. Jean's other son, Joseph Moliere Salvant Sr., struggled to become a small sugar planter in Jesuit Bend. Between 1850 and 1860 he increased his worth from $5,000 to $20,000, a not inconsiderable sum (the equivalent of about $125,000 to $500,000, respectively) though still only a tenth of the worth of the wealthiest planters in the parish. His wife Marie Josephine Veillon bore him eight children, including Grandmother's father.

Census data and marriage and probate records allow me to reconstruct in some detail the early history of my grandmother's family and of Louisiana's rural economy in the eighteenth and nineteenth centuries. This research has delivered me to a past that exists beyond stories of a family firmly rooted in New Orleans proper and the urban world of the Creole elite. Neither the few records nor our family stories lead to Plaquemines Parish. Oral history often acts this way. Time gets compressed in a process scholars refer to as "telescoping." What might be recounted as a generation, in fact traverses many decades and multiple generations. A far more complex and distant past gets simplified or even disappears within a single historical narrative that rests on a mythic charter, in our case the simple assertion (which is both true and false) that "we came to New Orleans from France."

Historians work to slow down and understand this foreshortening. I want to know something of the world of Plaquemines Parish, including the intimacies that unfolded among the Salvants and the human beings they possessed, like the seventeen-year-old Madeline who ran away from Jean Salvant III with a bounty of twenty dollars on her head. Slaves took the Salvant name. One mulatto slave gained his freedom and by 1860 became a slave owner himself. It's clear that miscegenation took place; one of Jean's sons would marry a mulatto woman. By the late nineteenth century black and mulatto Salvants lived and labored in Plaquemines and in New Orleans. Ultimately prejudice and exploitation pulled people apart; families scattered around the country tied by history, and sometimes blood, but mostly separated by race and oppression. There is some lineament that connects me to an African American professor of English teaching in the northeast, just as there is a shared history stretching across the Gulf of Mexico to an impoverished corner of Haiti.

The Civil War destroyed Louisiana's slave-based sugar industry, though Plaquemines already had seen its best days. The value of farms declined between 1850 and 1860 by nearly 40 percent. War brought a definitive end to the parish's economic fortunes. In April 1862, the Union Navy took forts Jackson and St. Philip near the mouth of the Mississippi in what proved to be a crucial moment in the war. As their men and ships moved north through Plaquemines to New Orleans, Northern forces met little resistance, giving them control of the Mississippi. Free blacks in the parish formed a militia to assist the Union.

New Orleans witnessed a rapid migration of people following the destruction of farms and the end of slavery. Many of the Salvants—white, black, and mulatto—moved to the city. Most joined the urban working class. One became a cattle driver, living with his mulatto wife Josephine. Those with some money used their knowledge of sugar to work in one of the city's merchant exchanges. A few prospered, such as Thomas J. Salvant and his cousin R. M. Salvant, who owned a large house at 3201 Ursulines. Others pursued different trades. One son of Joseph Moliere Salvant Sr. entered the retail shoe business, managing Dunn's shoe store.

For the fifty-year-old Joseph Moliere Sr., the Civil War ended his economic dreams. When war broke out in 1861 he had just entered the class of small planters in Jesuit Bend, wealthy enough to send his namesake away to a Jesuit school in Kentucky. Sometime around 1862 he left Plaquemines Parish for the city, along with the rest of his family. At the beginning of 1867 he purchased a modest single-story house in the city, on the corner of Columbus and Johnson Streets, and four lots of adjoining property. In 1870 he described himself as a "retail merchant." The value of his personal property stood at a mere $800 (about $12,000 today), a mere fraction of what he had possessed just a few years earlier. Josephine died in 1880 of cancer of the "womb." Joseph followed her fifteen years later. He left no will. A month after his passing, the seven surviving children petitioned the court to be recognized as heirs and to inherit the estate, which was then sold off.

His son Joseph Jr., my grandmother's father, began work as a retail grocer in 1870 with an estate valued at $500 (today just over $7,000). Six years later, he was laboring not as a cotton broker, as the family stories have it, but in the sugar trade. He had married the beautiful twenty-three-year-old Marie Antoinette Chevillon, whose French father worked as a carpenter and had taken a local girl as his wife. The couple first lived in the Third Ward, a relatively poor area of mulattos, free blacks, and immigrants, and the birthplace of Louis Armstrong. By 1882 they had moved into the Sixth Ward, the Tremé, renting a small house on Villere Street, where Antoinette gave birth to my grandmother, Mary Cecile Samuella Salvant, in late 1882.

The Tremé was home to former slaves, Haitian creoles, *gens de couleur*, and a few whites. At the time 3, 214 people lived there, only 176 of them white. During the antebellum period, the ward had the largest urban free black community in the South. It would become one of the most politically active and culturally rich communities in the country: the birthplace of jazz, home to black newspapers, and where African Americans protests over segregation in transportation would lead to one of the country's most momentous Supreme Court decisions, *Plessy v. Ferguson*, which upheld the "separate but equal" doctrine that made Jim Crow possible.

It was a rough, wild place, redolent with smells, sounds, and bacchanalian delights: jambalayas and étouffées, music tumbling down the streets, drunkenness and revelry. Grandmother's childhood coincided with the time when the Tremé became world-famous for its prostitutes in what became known as Storyville, the largest red-light district in North America. Across the street from their Tremé home sat one of the city's most famous brothels, at a time when virgins sold for as much as $800 and elite madams might make $140 a night. Through the day and well into the evening men visited "The Firm," where they drank and took their sexual pleasures. The Firm, went one advertisement, "is also noted for its selectness. You make no mistake in visiting The Firm. Everybody must be of some importance." Down the street men cavorted with "first class Octoroons . . . Any person out for fun among a lot of pretty Creole damsels, here is the place to have it." Kate Armstrong operated on Villere Street. She arrived in New Orleans a beautiful young woman with a particularly well-endowed bosom, and became one of the city's most sought-after strumpets. Armstrong ran a renowned brothel furnished with marble tables, pianos, damask curtains, expensive oil paintings and gilt-framed French mirrors. Around the corner in some of the city's roughest areas were the lesser-known brothels and cribs, where some prostitutes practiced erotic voodoo ceremonies and accommodated anyone who had a few coins in his pocket.

A stone's throw from the rented house sat Congo Square, where on Sundays slaves had once danced and played on log drums, mule's jawbones, kalimbas, and xylophones. Grandmother grew up surrounded by people in the streets performing African dances and the improvisational sounds that would become known as jazz, songs like "Bucktown Slow Drag" and "Te-Na-Na" ("Little Nothing") by musicians who laid it on thick with their drums, trombones, trumpets, and saxophones.

> Take that pork chop to your Pa
> Take that sausage to your Ma
> Mommy's got a baby
> Te-na-na, te-na-na.
> I've got a girl living on the hill
> If she won't, her sister will
> Te-na-na, te-na-na.

Joseph and Antoinette married at St. Augustine Church with its Ursuline nuns and pews arranged in a circular pattern around the altar in the middle. Slaves and *gens de couleur libres* had attended the church, the liberated slaves purchasing pews so that blacks and whites sat close to one another. The couple prayed alongside Homer Plessy and other members of the black elite. To more and more whites, the ending of slavery had created too much wildness. Blacks began shouting during sermons. Others fell to the ground, writhing before the sacrament. Emancipation erased the ability of whites to enforce separation. My great-grandparents were among the disenchanted, and the family moved to the German Chapel on the corner of Galvez and Laharpe, outside the Tremé and across Esplanade, in the better—and whiter—part of town.

My grandmother spent her childhood in the city's violent red-light district when white and black Salvants lived near each other in the fabulously racially complicated city. Over the course of Joseph's married life the family lived in eight or more houses, all in the Tremé. Grandmother never told anyone of growing up amongst the blacks, prostitutes, street musicians, dockworkers, gamblers, and petty crooks, perhaps because her own parents disavowed the city's rough-and-tumble bacchanalia. Racism would divide people who shared the same last name and, in so many cases, the same blood. Those who could became definitively white. Connections to Africa and the rich history of living in Plaquemines disappeared. Over the decades, living in one shotgun after another, Grandmother created a story of herself and her family redolent in details of vanished wealth and pure French ancestry. The stories became true in the telling. She believed them. Cecile believed them. So has everyone else. We were once rich, New Orleans aristocrats, upstanding members of the Creole elite in our Esplanade house tended by servants.

In the fall of 1891 a horse-drawn cart carried the body of forty-four-year-old Antoinette across town from the house near Congo Square to St. Louis cemetery, where she was laid to rest in the Chevillon tomb next to gamblers, pirates, politicians, and voodoo priestesses. Grandmother was nine. Her oldest sister Stella tended the house. A few years later, Joseph Moliere Sr.'s estate provided some income for his children. Joseph Jr. soon thereafter married the young Mary Louise Gamotis, the daughter of a well-established apothecary owner who had emigrated from France. They married very near the sixth anniversary of Antoinette's death. Mary was in her early twenties, half the age of her husband and not much older than Grandmother. The couple had one child, Emilie, born in late 1898, near Congo Square.

Joseph Jr. finally escaped the Tremé to a sizeable home a few blocks from Esplanade and near where his father had once lived. He had worked hard to improve his fortunes in the years after the Civil War, using his family's earlier history in the sugar business. By the time he married his second wife, Joseph Jr. was working as a broker for one of the many industrial refineries that processed the cane along the Mississippi River. He took an active part in the Catholic Church and had become an upstanding member of the white New Orleans community. He finally owned a property he could call his own. The following year, in his mid-fifties, he died, his youngest child barely a year old. Grandmother was seventeen.

Under Louisiana's rococo inheritance laws, Mary controlled the estate, which passed, upon her death many years later, to her daughter and only child Emilie. The nine surviving children of Joseph Salvant Jr. and Marie Antoinette Chevillon, who had grown up in the working-class and racially mixed Tremé, inherited nothing. Not long after his death Grandmother and the other children left their father's house and moved in with one of her older brothers, renting a house a few blocks from Storyville. She soon began work as a teacher. Emilie died in 1968, leaving an estate valued at $7,565, which Grandmother inherited as the sole survivor of her parents, though the estate's dispensation still remained unresolved when she suffered a fatal stroke three years later.

—⁓—

As I walk past the faded resplendence of Esplanade, I wonder what forces sustained the making of familial myth and the silencing of history. Grandmother, who loved her father, seldom if ever spoke of her stepmother or stepsister. I suspect that the appearance of this new young woman in her father's life was unsettling, and that his death and the disposition of the estate estranged Mary from her stepchildren. I can conjure an image of Grandmother packing her bags and leaving her father's house for yet another rental in yet another rough part of New Orleans. Did this tumult in my grandmother's life—the unexpected death of a beloved parent, economic distress when she had only recently tasted prosperity— somehow sustain desires that grew into fantasy, including the archetypal stepmother who steals one's birthright?

Her father had actually achieved some success, though neither Joseph Jr. nor anyone else ever lived on Esplanade. Perhaps this brief period was enough to nurture rich imaginings of a lost world that might return any day, a past purloined. Perhaps she inherited her father's pearlescent dreams born in an Antebellum south of sugar and slavery, dreams his family had just begun realizing when Union forces made their way up the Mississippi.

Grandmother came of age during an important moment of historical myth-making in New Orleans. The era of white Creole dominance was coming to an end. Affluence moved away from the historic center of the city to the grand homes of the new elite along Saint Charles Avenue. In some respects New Orleans was becoming a more American city, and more racially intolerant. The din of prejudice made it difficult if not impossible to discuss the complexities of race and slavery in Louisiana and urban life in New Orleans. White Creoles invented narratives that simultaneously regaled their French heritage and denied the mixing that so powerfully shaped the region's history.

The archive reveals little of the interior worlds of my grandmother and her family, and nothing to explain my aunt's dreams and the still powerful insistence by family members of lost wealth and refined descent. The evidence that composes a reliquary of lost times—the records of births, deaths, and estates, the forms completed by census officials, the documentary minutiae that fill countless repositories—still beckons, the ceaseless yearning to speak with ghosts. History, like memory, is time travel. We bump into others and into our selves, and yet they are never quite our selves, never quite the other.

THREE

—w—

SECOND STREET

KINTA BEGINS CRYING BEFORE I'VE EVEN BEGUN THE interview. Each word brings a new tear, etching mascara lines across her face.

"There was no touching. All of our lives there was no touching. Mother can't touch. Hugging is real awkward . . . you don't even want to try."

I stop the recorder, walk to the kitchen for some tissues and a glass of water. Drinking water helps you stop crying, my sister Marie always says. "C'mon, c'mon, drink another glass," she likes to say. "Ya gotta drink water, lots and lots of water, when you cry. It'll help ya stop." But as I walk back to the living room, and Marie reaches for another whisky, I wonder if I am not simply replenishing some hidden aquifer of sorrow. Kinta will drink from the glass and never stop crying.

My sister Sabrina also remembers the absence of maternal touching. Mom cooked meals, stitched clothes, and of course certainly held us, but she generally avoided physical contact, at least from about the mid-1950s, when she abandoned her dreams of financial security for a bottle of Bourbon and a pack of Salems. A decade earlier there were Mardi Gras costumes and birthday parties and stories shared around a meal of roast chicken on early Sunday evenings. But with each day it seemed as if her dreams were slipping through her fingers. She would never stop moving, never own a house, and never cease worrying about the bills that came each month. By the time Sabrina was born in 1955, Mom was drinking heavily. Children she had once welcomed into her life now seemed like burdens, unwelcomed if not yet disavowed.

During the 1950s the expected motions of family life dissolved. Mom and Dad still sat at the table, but miles could have separated them. Dad did not want to be there; he was in his forties, living month to month, and by the end of the decade he would be faced with a lawsuit for failing to pay his debts. Dad would try remaining still, indefatigably still, as Mom railed against him, calling him a coward, a failure, the invective turning to a boozy slur until one day he got up and walked away. Dishes stacked up in the sink, children were left unsupervised, and Mom retreated to her bottle and bed.

We still try reaching out to her, hoping for some physical connection, perhaps even comfort, a fantasy that at some moment she will return to being a mother. "Hi Mom," we say, as we bring our arms over her shrinking body. But she never puts her arms around us. There's no reciprocating affection. She just stands there. It's as if we are holding onto an apparition.

I don't know why Kinta began this way, and I don't

ask why sitting down at an appointed time to discuss the past suddenly unleashed such emotion. I didn't intend on starting the interview by trying to reconstruct my sister's relationship with Mom, or expect that she would want to talk about something as simple as a mother's touch. I sense Kinta didn't mean to start this way either, though I wonder if her taciturn demeanor doesn't hide a torrent of emotion just beneath the surface.

Historical research often unfolds surreptitiously. You never know what you're going to find when you open up an archival file or begin speaking with someone. The historian's professional image suggests something different—that we know exactly what we are doing as we labor to master a particular past. In fact, research is rather messier work, which is one reason we feel the need to have methodologies, some way to order the chaos, some way to organize our disquiet. And yet it's in the unexpected things and furtive glimpses into people's entangled lives that we sometimes discover what is really important.

This sense of the past as here and not here is at the center of our consciousness and our historical imagination, which may explain why behind all history lies philosophy. Historians may begin with a few general questions, but they usually end up confronting larger issues: What is the nature of memory and its relationship to history? What are the possibilities and limits of our understanding about some distant time, a century ago, or just yesterday? Does the past in some basic way remain forever hidden and unknowable? What is truly past? What remains still present?

Over the course of our lives we ask similar questions of ourselves. We wonder about the forces that have shaped who we have become. We worry about what might have been. We search for explanations, try identifying the extent to which we control our destinies. What is my relationship to a friend, family member, lover, to my ancestors? We consider our own histories, realizing that in all memory resides the pasts of others.

—⁓—

Sitting down together, turning on the digital recorder, knowing that I am searching for a past I can't remember, has unleashed a flood of memory in my sister. I expect Kinta to continue talking about touch and maternal attachment. I am ready to offer what I've discovered about memory, neurotransmitters, the look of an eye. But she changes the subject from absence to a life-threatening event. I wonder if her mind hasn't somehow formed an association between her conflicted attachment to Mom and a sense of being surrounded by ever-present danger.

"Do you remember that man? I think he was a sailor. You don't remember him?"

I shake my head and scan my mind, hoping that talking to my sister will somehow jar something loose. But I have no idea what she's talking about.

Kinta starts telling me a story from when she was about fifteen and I was seven or eight. She explains how clear the memory is, as if it had happened yesterday, though there are details she doesn't want to say, or doesn't herself know. Did she open the door on the man having sex with our mother, or did she hear him beating her up, and that's why she asked him to leave?

"'You got to leave,' I said. And Mom was passed out. I mean drunk. Incoherent, in her slip, lying across the bed. And for some reason he left. And for some reason, I don't know, I locked the door."

She pauses.

"We always left the front door open, but something told me to lock the door."

Kinta looks down for a moment. She seems a bit older now, a bit faded.

"And he came back. He came back.

"'I left my wallet in your mother's room,' the man said.

"'Well, you'll have to come back tomorrow and get it.'

"But the sailor needed his wallet. 'Open the door, Goddammit.'

"'I'll go back and look for it,' I told him. And I went and looked, and I couldn't find it. I even looked under the bed. Mom was just lying there, passed out. And something told me not to unlock that door. And he left. And he never came back. He never came back."

"You don't remember?" she asks me again, hoping the story has jogged my memory. She seems amazed I have no recollection of this threatening man in our house.

"No, I have no memory of this at all. But I have only very few memories of childhood." I reel off the half-dozen memories scattered across my mind. Most often they feel more like sensations or feelings, barely discernable images devoid of context.

Kinta is surprised. I only told her about my memory problems when I began digging into my own past and the origins of forgetting, though each time we talk she seems perplexed anew, as if it is the first time that I have shared the fact that much of my childhood has simply fallen into some sort of abyss. How could I not remember, she wonders? Kinta's worry has been that I have too much memory, not too little, that her "baby" brother was scarred by an excess of experience. She imagines me as overwhelmed by memory; it seems only natural that I became a historian.

Her mind gushes with memory, going back all the way to when she was a little girl living outside Houston and Mom found a rattlesnake sunning itself by the kitchen door. Kinta can reel off one story after another, little histories really, of Houston, Mississippi, New Orleans, dressing up dolls, make-believe games, childhood adventures, the lovingly immutable relationship with her younger sister, Sabrina. She finds solace in memory and sisterhood stories. Sibling relationships become narratives of comradeship and solidarity in those years when everything else seemed to be falling apart.

I envy Kinta. She has memory to spare, like the memoirs I've been reading where the past is as clear as the time it happened. She has what neuropsychologists call an exquisite episodic memory, the perception of "things past" and the ability to bring into conscious awareness moments that define a life. My mind seems nearly blank in comparison, like a sheet of crumpled-up white paper.

Part of me wants to turn didactic, professorial. Mom's alcoholism and depression, her anger at *everyone*, shaped the bond between mother and child. Certainly Mom touched you and held you, even if you can't recall an instance. An image of the past stands as a story, the absence of touch a metaphor for an absent mother. My sister's feeling is not a fiction, even if it is not entirely true. She knew then, as she knows now, that in the 1950s our mother began retreating from her role as a caregiver as her life collapsed into disorder and despair. All these stories, this surfeit of memory, are not facsimiles of some objective reality. They are important precisely because they are stories. Each is a small history of her self.

—⟋⟋⟍—

A past exists in the simple statement "I am." Personhood has its own particular history. Children begin developing a sense that they are someone around the age of five. At nearly the same time, they form an idea that time moves irrepressibly forward. They become little historians and memoirists, telling stories of cause and effect, of who they are, and how the world came to be, however fanciful their imaginations. We come to live by these images of the past.

Before this entrance into a world of time and becoming, the boundaries between self and other remain indistinct. There is no time, just a constant state of being, in which the baby remains a part of the mother. The child lives inextricably tied to the world, awash with sensation and free from time's arrow. The joys and sorrows of others, including and perhaps most importantly the parents, are their own. We are one with our mother or caregivers, no matter who they are. These early sensibilities may be related to the functioning of what scientists have described as "mirror neurons," which reproduce or simulate actions taking place in another organism—a smile, a tear, a funny face. The neural basis of imitation and empathy remains unclear, and research on mirror neurons is very controversial. What seems more certain is that the baby's brain acts as if the performance of someone else's body is its own. We begin our lives a part of others in a state of timelessness. This may be why the feeling of empathy seems to suspend time; identifying with someone's emotions or their situation returns us to that original oneness.

Order unfolds as we seek relationships, beginning with simple imitation. Patterns materialize. Mental pictures arise. Repetition helps create the long-term memories that become the repository from which we tell the story of our lives, assigning importance to particular recollections while demoting others.

Becoming a person, however, means being alone as well as living amongst others. We know something of how this process unfolds within a child's brain. There is a burst of neuronal development followed by a pruning. The brain cuts out what it considers unused and unimportant. An age of exuberance recedes like the afterglow of an exploding star, present if invisible to our conscious senses.

Within certain limitations, the brain is historically constituted, the external world shaping the development and relationships among and between neurons. In a sense our brains are like an archive, where material is well-preserved and properly catalogued, but also dissolves, or becomes re-shelved or misplaced, or in some cases never makes it there to begin with. Time is deposited within us, as the brain is also constantly interpreting the external world. This historically created brain and the self emerge in tandem. We tend to think of the self as something concrete and fixed and not as a process emerging from our awareness of experience. We want to say that the self is a kind of private property that we, and only we, own, so we can point to someplace in our brains and say the self resides "there."

There is a relationship between this historically made brain and the self as a story unfolding along an axis of time. Our sense of being emerges as if creating a story, much as a historian fashions a coherent narrative explaining a process or development. Through history telling, in the words we begin using, the child integrates experience, consolidates memory, and develops a consciousness of self, even if that self in some respects is always a fiction. Like language, the self struggles to prove its truth, marching toward a point forever disappearing in the distance.

We now have a sense of the neurological basis of the self. Neuroscientists interested in memory's creation have focused considerable attention on the amygdalae and hippocampi. The amygdalae play an important role in storing stressful moments and are related to instinctual memory around danger and fear. The hippocampi, more newly evolved structures within the brain, allow us to fix experiences in place and time. They are central to the creation of declarative and autobiographical memory, the ability to be one's own storyteller, a historian of our selves. Without the hippocampi there can be no self, no society, no culture, no history-telling. The two structures operate in tandem. Chemicals like cortisol, adrenaline, and noradrenaline cascade through the amygdalae and hippocampi and across our brains, particularly the prefrontal cortex and the language centers, enabling me to type these words and Virginia Woolf and countless artists and writers to explore the paradoxes of our being.

Powerful or traumatic events can alter or even hijack this process. Traumatic memories stored in the amygdalae produce bodily experience and shape perception, from the sweat that automatically appears, to the chemicals rushing through my body causing panic whenever I see my mother. Should I run away every time I see her, every time I feel her presence within me? Trauma obliterates time. The memory is never past. My mother's suicide attempt remains timeless amidst all my forgetting. This is because stress hormones mold the way we create vivid and rich pictures of our lives. Trauma trips up the elaborate choreography of being, even as that single event dances on and on. The result is one kind of amnesia, reducing everything around that original experience to gray haze, as if nothing else happened. For the soldier with Post Traumatic Stress Disorder (PTSD), the war never ends.

Severe trauma quite literally destroys parts of the hippocampi, making it difficult for someone to tell the story of their lives. Depression does much the same thing, shrinking the hippocampi and altering the ways we remember, describe, and interact with the world. At its insidious worst, depression destroys the self, leaving one with the sense of simultaneously being devoured and disavowed by one's past. Writers and artists have depicted depression as a kind of dissolution, as if their minds were literally breaking apart, as if their selves were somehow slipping away. The paradox of the self—identifiable yet impermanent, nonexistent though obviously there—may also explain why so much art, particularly the modern novel, has been preoccupied with how we become who we are.

Childhood trauma related to chronic amnesia seems to involve experiences that simply overrun the child's ability to order or assimilate the external world. These experiences typically are repeated over the course of early lives, in addition to punctuated extraordinary events, and they shape the evolving brain. The hippocampi atrophy. The past possesses these people because they cannot yet put into words what, as a child, overwhelmed them, nor can they summon the richly detailed world within which their young lives unfolded. Memory is both insistent and missing or subsumed, an "impossible history."

I became myself around the time my mother wanted to die. I saw in her eyes anguish and despair, emotions I considered my own. One consequence was childhood depression. At least that's what people have told me.

"You were a very sad child."

"You looked kind of abandoned, like you were walking around in state of shock."

"You were in trouble."

"You had PTSD."

Or "these forms of stimulation are unhealthy for the young brain."

Or "as a child it was impossible to understand everything that was happening around you, which you couldn't control, couldn't possibly understand as a young child. Your mother was an alcoholic and in such crisis. And you were surrounded by adult sexuality, your mother's cravings, the men, your sisters . . . [and] you so wanted your mother's love. And you were so disappointed."

Or "it appears there's nothing there. The slot inside your brain is empty."

I still dream that I will tell a story of my self that contains the full richness of experience—a moving picture instead of a few scattered images. I will take from others enough information to create memories of boyhood dramas, school scenes, adventures, plans about what I would be when I grew up, relationships with parents and siblings. I will mold collective memory into personal testimony. Language will fill in the spaces where my body has been.

I can't stop hoping that there are memories inside of me that somehow are unavailable but that with enough effort will be revealed like a sunken ship brought to the surface. The past will become incarnate, though I also tell myself it is forever lost. I am wasting my time trying to figure out something, anything, and this appointment with the past may well destroy me. I will show up, but no one will be there. I will sit down and wait and slowly waste away.

I think for a moment about the play *Oedipus Rex*, wondering if Freud had it all wrong. The play's not about sex or incest, or at least not only. It's a history lesson, a cautionary tale. Oedipus wants to find out what happened, why an as yet unknowable past is determining his life. He's on a quest for knowledge, the facts that will finally put everything in order, the continuities that comprise human tragedy.

"Don't do it," the Muses warn Oedipus, "this is a history that is beyond comprehension, beyond language itself."

"I have to," Oedipus insists. "I must know. I must transform the past into memory."

"No, Oedipus, this is man's fatal desire, his wish to order the unmasterable. What you want exceeds the capacity of irresolute mortals. You won't be able to bear witness

to your past. There is no amnesty in memory, only the sham-
ing recognition of man's flawed being."

It's not a crime that destroys Oedipus, or some broken
taboo. Oedipus didn't know that he'd slept with his mother
or killed his father. It is knowledge that blinds Oedipus. Putting things in order, assembling all the evidence along time's
stubborn arc, finally telling his impossible history, is simply
too much.

Kinta's voice draws me out of my head.

"I fantasized all the time," she says—dreams of finding
riches buried in the backyard, wealth that would make
amends, men who would be kings. She would live with Dad
on the Gulf Coast, just the two of them. They would fish
off the Bay St. Louis Bridge, dipping cane poles into the
water looking for croakers, trout, and sheepshead, pulling
up nets filled with blue crab, drinking ice-cold Cokes from
a Styrofoam chest. In summer evenings they would wade
through the warm shallows looking for flounder beneath
the custard-yellow light of a Coleman lantern, the tide running around their legs.

"Sometimes I just don't know what's a fantasy and
what's a memory. You know . . . a real memory." Kinta
looks at me, hoping my research might help her arrange the
jumble of sensations that course through her.

"I listened to so many stories. They feel like I *really*
experienced them. I can't tell if I heard them, or if I was
really there, part of the story."

A worry befalls her. There is something dangerous in

this work of memory. The compulsion to unravel what seems authentic from what might be fantasy leaves her feeling exposed, even diminished. She seems to be beholden to some impossible task, obeying some ancient injunction impelling her to organize all these feelings and recollections, to separate a daughter's unrequited dreams of a perfect father from sitting together on a bridge watching a float bob up and down on bronzed waters.

It's a Sisyphean task.

"Maybe the stories you heard as a child and your fantasies, and what you really witnessed and today remember, aren't so far apart," I am tempted to say between her tears and sips of ice water. "And our dreams are also the facts of our lives, every bit as true and often more powerful than the empirical events we think we can so easily document." But this would come across as hurtful, or patronizing, or both. And I wonder if I'm not being hypocritical. It's the facts I am looking for after all, the concrete evidence I keep telling myself will somehow put things in order.

Kinta switches the topic slightly, telling me how she had to escape from New Orleans, even if it meant leaving us behind. She seems to want to talk about sexuality, her own and our mother's. It seems a way of speaking to her feelings of vulnerability in those years around the divorce, when Kinta was trying to make sense of herself and the desires of men, and our mother's sexual hungers and endless disappointments. Kinta knew she was headed down the wrong road, staying out all night, hitchhiking around the city, making out with boys who drove too fast and boasted too quickly about drinking nickel beers at Two Jays on Governor Nicholls. It was just a matter of time before she would lose her virginity and, soon after that, become pregnant, Marie warned her—Kinta would end up like her or worse, with a baby hanging from her hip, the toilet filled with dirty nappies, the loneliness of young motherhood.

She was only thirteen then, with tanned skin and hair as dark and shiny as anthracite that ended in fishhook curls that hung around a child-woman's body. She'd developed breasts early and had a preternatural beauty that made her seem older and worldlier than her years. Her looks drove boys wild. Kinta liked necking, and she French-kissed until the muscles in her mouth grew sore. Sometimes she would let their hands sneak under her top. But they were all only boys, she knew, with their cars and music and doggedness—none promised much of a future.

Running around New Orleans at all hours of the night had its thrills, and it kept Kinta away from our mother. By the time she got home Mom would be in her slip passed out on the bed. Kinta would make herself a sandwich and slide into her room. But my sister knew that everything was happening too soon, too quickly. She needed some direction, some escape from the chaos. She needed a parent. A few months after the divorce was finalized Kinta got on the Greyhound bus and joined Dad in Mississippi.

They lived in a trailer at the back of a house in Gulfport that belonged to Pat, the woman our father was seeing. Dad had met Pat through my oldest sister Susan, who dated and then married Pat's son Joe. Pat had been a vaudeville performer from southern England. During the 1930s she toured Germany while the Nazis rose to power. She came to America with her husband; they were on their way to Hollywood to make it big in the "talkies," they assured themselves, leaving the dreariness of bombed-out London for Los Angeles's promised light. But Pat's husband died. She never made it out West. With the few dollars she had in her purse she bought a little bit of property in Mississippi next to the train tracks, took a job as a telephone operator for Southern Bell in Gulfport, and showed anyone who asked how to Charleston.

No one talks much about this period in our father's life, including Joe and Susan. My other siblings aren't able to offer much information. Divorce and familial crisis jumbles the expected flow of time, making it hard to assemble an ordered story of anyone's life. But I wonder if behind the silences lay something else. Postmarks and legal correspondence reveal my father living in Mississippi around the time of the divorce, and that he was already involved with Pat. Their relationship had begun in the late 1950s when my parents lived on the Gulf Coast and Joe and Susan had begun dating. Mom was drinking heavily. She also had grown attracted to a younger man, a teacher in an English class she took at night, though the experience left her feeling bereft. Nothing but an alcoholic rage could flow into the spaces deep inside her that had been hollowed out by years of taking care of children and a marriage that never lived up to its early promise. Mom railed at Dad's long history of broken employment and his saccharine pledges of a better job next time. Dad drifted away. I do not know when he and Pat started sleeping together, but it seems clear that our father was having an affair while our mother's life was falling apart.

I visited once for a short while during the school year. The documentary record tells me I was in first grade. I want to say it was the fall, perhaps October, after the Southern heat had broken and the light had become a littler clearer, a bit more pure. It was a few months after my father had pleaded with the court to grant him custody. Kinta tells me we slept in the trailer. I try imagining what the days were like. Kinta was at St. Joseph's Catholic school in Gulfport, and the adults were at work, so I was alone most of the day. Decades later I stand watching diesel engines with their tanks of toxic chemicals rumbling through the stands of pine trees. The smell of the sulphurous rocks conjures an image of my waiting for the trains to pass, sitting on the gleaming tracks that were nailed to wooden ties blackened with creosote. For a moment I can see my childhood self balancing a penny on the track, looking for it after the brown caboose disappeared into the pine trees, finding it flattened smooth and shiny.

—⁂—

I am driving along the Mississippi Gulf Coast, heading east from New Orleans. Highway 10 whisks you through the northern parts of Bay St. Louis, Pass Christian, Gulfport, and Biloxi. Billboards summon drivers to the casinos. There are concerts to hear, food to eat, and, most important, money to win. A woman smiles down at me. "Letisha won $20,000. Next Exit." Strip malls turn into a concrete and plastic blur before the interstate dips into a forest of pine and swamp.

Most everything along the highway seems perpetually new, as if the architects and urban planners figured out a way to erase history. It is now impossible to tell that Hurricane Katrina devastated this area. To see the hurricane's destruction you have to go inland a few miles to Old Route 90, known to locals as Beach Drive. The road began in the colonial period as a small trail, connecting the various coastal towns whose inhabitants mostly lived by scraping the sea bottom for oysters or dipping nets into the Gulf for shrimp. The storm leaped across the road and tore down one house after another, leaving behind front stairs and foundation pads. A decade later you can still see heavier personal items peeking through the dirt, a spoon or a pot or some other shard from someone's life.

In the years after the storm the Biloxi county government worked out of a trailer. The public library was in a trailer too. The storm exacted a terrible human toll. Families cracked apart. The divorce rate skyrocketed. The death rate increased as well. Quite a number of the elderly simply gave up living, as if they had said, Enough is enough. The casinos were the first to return, rebuilt far more quickly than homes. I drive up to one casino, read the billboard for the next concert. There is a competition to win a brand new house. I wonder if any of those displaced by the hurricane thought of taking the chance.

I cross over the St. Louis Bay Bridge and drive through Pass Christian. A few people have parked their cars and are playing in the water. The sand is a radiant white, the gray flat waters of the Gulf of Mexico ghostly still. I take a left turn, drive a few blocks through the ruins heading east on Second Street, looking for the old house, hungry to remember that summer of 1967 when I joined my father and siblings in Mississippi.

Nothing comes of these wanderings. Even when I think I have found the house, doubt creeps in, and no new memories follow. Nonetheless, I take photographs and jot down a few notes, adding them to the archive of my self, hoping that by some magical transubstantiation they will turn into personal memory. It's an article of faith we amnesiacs have. By chance or by effort our pasts will return. The gap will be closed between what is ours but lost and the world that holds our traces. This archive will be my remedy, I tell myself, the place outside me that will fill in the empty spaces.

There is enough information to get a sense of my father's life in Mississippi in the mid-1960s and to imagine the time my siblings and I spent there. My father found a new good job, perhaps the best job he ever held, at the Stennis Space Center in Hancock County on the other side of the bay. In April 1966, NASA had begun the first static test-firing of the Saturn V rockets that would take Neil Armstrong and Buzz Aldrin to the moon. NASA tested the rockets late at night or in the early morning hours before sending them along Highway 10 on the slow ride east to Kennedy Space Center. I have an image—not a memory, more of a feeling, even a fantasy—of the rocket engines strapped down like some primordial beasts and the earth shuddering against their fury. Outside my bedroom, branches sway and pecans tumble to the ground and heroes whirl through the cosmos.

The Space Age brought good steady jobs to an area that mostly relied on fishing, tourism, and Keesler Air Force base in Biloxi. Dad worked out of a trailer doing the books for a NASA contractor. He was single again, divorced from

Mom, and no longer shacking up with Pat. He found a small rental house a half a mile from the beach and set back from the road on a huge plot of land filled with pecan trees and dappled light and shadows that seemed to go on forever. From the nearby pine forest marched the evening chitt-chitt-chitt of crickets, and in the summer, after a thunderstorm had cleaned the air, fireflies turned on and off, obeying some inscrutable commandment. There was a kitchen in the back, and in the front a generous screened-in porch in which to while away the time. The house was clean. Once a week a black woman swept and mopped, hung out the laundry, made the place neat and ordered. We had an old three-legged dog, Chester, that chased squirrels and one day disappeared into the forest.

I like to think that for a while at least we were children again, or perhaps children finally, and that the world scarcely extended beyond the beach a few blocks away. Gus had managed to go part of the way through high school in New Orleans, living on his own in a beat-up apartment. Mississippi offered a reprieve of sorts. He hid himself in broken-down cars suspended on cinderblocks and surrounded by Craftsman wrenches and sockets scattered on the ground like the bones of some torn-apart carcass. When Gus wasn't working on a car he was out on the Gulf fishing, returning at twilight sunburned and content. My brother was happy out on the water and in the evenings driving fast on Mississippi's straight roads, or at least happier than he had been for a very long time. He could try forgetting for a moment what he had endured in New Orleans and the war that was taking one friend after another to Vietnam.

In 1967 Kinta was fifteen, with an innocent face and a

body that suggested an experienced woman. In the summer months she headed to the beach, unfurled a blanket and coated herself in Coppertone cocoa oil. She turned dark and lustrous. Once she fell asleep on her raft and floated out with the tide more than a mile before some fishermen rescued her. She was a beauty queen and a cheerleader. Boys simply couldn't stay away from her. She spent a lot of time on the beach, making out around bonfires—good, clean, late-1960s teenage fun, she tells me now, kissing, letting boys feel her breasts, sometimes even heavy petting, but never real sex. Dad would be asleep in the front room when she got home from a date. Kinta would tiptoe past him, change the clock from midnight or one o'clock in the morning to ten-thirty, and, once he went to bed, sneak in again and return the clock to the proper time.

Kinta and Sabrina always had been close, but in Mississippi they could be lazy, lounging out on the beach, going to the pool, whiling away time, a hiatus of sisterhood, for Kinta a final pause before womanhood.

Kinta often reminds me that she taught me how to swim; she says it so often that I have created a memory of what happened. Or is it a fantasy?

Go ahead. Jump in. Come on. I'm right here. I'll get ya.

I am standing at the edge of the pool. I can see Kinta's legs kicking outward. Her hair is long, black, the ends curling in the bluish water. She is smiling and teasing me, but only gently.

Come on. I'm right here. I'll get ya. Jump.

I feel the water, clean and cool against the sun on my back in summer's still heat. I try to keep my head above the water, but I feel myself going under. Then two hands grab hold of my chest and push me upward. Panic turns to glee. Kinta brings me to the edge so I can climb out, do it again. I do this until my knees are raw, climbing out, running around, jumping, being caught, until I realize I've not paid attention. I am in mid-air, and Kinta isn't there. I tumble into the water. I look up. Light shimmers, images split into a thousand pieces, until my head breaks the surface. I flail at the water but manage to move, until I reach the edge, pull myself out, realize I can swim.

On weekends we went fishing and crabbing. Dad bought cane poles, crab traps, and chicken necks and shrimp for bait. We drove out to the Bay St. Louis Bridge. We mostly caught catfish and croakers and the occasional trout, but sometimes we came home with enough crabs to boil in a pot seasoned with a bag of Zatarain's. In the evening we would sit out on the porch picking crabs and eating cold watermelon sprinkled with salt, listening to the crickets and watching the fireflies.

My sisters sprayed Aqua Net into their hair and talked about boys. Gus had become a man who worried about a faraway war and carried the injuries of being an adolescent amidst our mother's insanity. I was the odd kid out, but in Mississippi it was the closest I would ever be with my siblings. We were children again, for one last time. Pass Christian seemed eternal. The Gulf's muddy salt air drifted through the trees and settled on our porch. Raindrops tapped on the metal roof. We collected pecans and spent hours on the porch cracking nuts and chasing one another through the house, slamming the screen door, squealing with laughter. The white badminton cock sailed through Sunday afternoon skies.

I came down with a bad case of measles. I don't remember when it was exactly, but it was sometime in winter when men were out hunting rabbit, squirrel, and duck, when on clear days the sky turns a cotton candy blue. I took the big front room where my father usually slept. The blinds were pulled, the room darkened. From the windows and the door that my father kept closed, thin lines of light turned the old yellow pine floor to shades of deep golden brown.

Sometime that week my mother arrived. She placed a cool compress across my forehead. I imagine her sitting there in the darkened room in a chair near the wall. At times I think I wondered if she was even really there, or if I was dreaming. I do not know where she slept the night, nor did I ever hear my mother and father talking, but I was faintly aware that somehow my parents were together, and that they shared a concern for me, their son.

—⁓—

It must have been in the spring of 1967, and it must have gone something like this. They arrived straight from the courthouse. *This is Dorothy, my new wife*, Dad announced. Dorothy smiled, cordially, went to the bedroom, and unpacked her bags. It was as simple as that. This is how things were with Dad. He lived by quiet proclamation.

Dad met her through Pat, the woman he had been seeing around the time of my parent's divorce. Jeanette, Pat's daughter, was married with two daughters of her own. Dorothy was Jeanette's mother-in-law. Jeanette drowned when her car tumbled from one of the bridges between Gulfport and Bay St. Louis. Dad went to Jeanette's funeral. Dorothy was there, and so were my sister and her husband.

There was a reception at Pat's house, where Dad met Dorothy for the first time. They started dating soon after and married within a few months. The divorce came two years later.

Dorothy despised the house. It was old and dirty, too much in the woods, too uncivilized. We moved to a small subdivision away from the beach, a typical 1960s planned neighborhood of nondescript ranch houses deposited alongside a cement road that curled about for no particular reason. There was a clothesline on the side and a chain-link fence enclosing a flat square of grass, and inside there were suburban amenities like air-conditioning, Formica counters, and a low minimalist couch and glass coffee table opposite a television.

My sisters called her "Dorothy the Witch." Dorothy got rid of us one at a time. Gus went first. I was next. Dorothy considered me repellant, deviant. Evidently I played with myself. "He has to leave, now," she told Dad. By September I was back in New Orleans. Kinta she both envied and detested. Dorothy accused Kinta of being lascivious, but when my sister was not around Dorothy tried on her bikinis, especially the pink one. Dorothy would sun herself in the small backyard, wearing the bikini as if it might somehow replenish her youth, and in the evenings she giggled and laughed as she and my father made love in the shower. By Thanksgiving, Kinta had returned to New Orleans. Only Sabrina, then just becoming a teenager, remained behind. Soon the three of them headed west for Southern California.

"It was like the end of an era," Kinta says. She had been away from New Orleans for nearly three years living with our father, often just the two of them, a child's dream come true. She had thought he would be like the prince of her childhood dreams, providing the protection she had never had in New Orleans—a parental shield strong enough to allow her to venture safely into adulthood. Now another woman had come between them, and worse, had tried to become her. Kinta could imagine Dorothy peeling off her pink bikini and frolicking with Dad in the shower.

Kinta didn't want to return to New Orleans. She loved Mississippi, the warm Gulf upon which she floated away the summer days. Kinta had just begun tenth grade, not yet sixteen, and was a basketball cheerleader. A boy was crazy about her. But there was no choice. Dorothy hated her. Kinta fastened a dress to the inside of her jacket, stuffed pockets with underwear. Her boyfriend drove her to the Greyhound Bus station. Kinta was back in New Orleans by the end of the day, and Dad did nothing to bring her back.

Kinta entered Fortier High School and took a job downtown at Maison Blanche. Work and school and friends kept her away from the apartment, kept her away from Mom's boozy ranting and all the jealousies she hurled at her daughter. Kinta left the apartment as soon as she graduated, enrolling at the University of New Orleans and working a few nights a week at the Playboy Club in the French Quarter. Management made her pay for the white cuffs, the black bow tie, and the blue satin outfit. In the dressing room Kinta sat with the other bunnies in their black stockings, curling hair, smoothing make-up, and drawing their Maybelline eyes.

Kinta liked exciting the men, protected by the strict rule that they could look but not touch. She learned how to do the bunny dip while serving drinks so that customers could enjoy her young body as they traced the dark line between her breasts. She watched their eyes turn liquid and wistful and hungry. The tips they offered brought her closer. She could smell the tobacco and the sweet Manhattans on their breath, and see their eyes falling across her olive skin. The Bunny Mother promoted Kinta to a costume of red satin and to playing bumper pool in the Playmate bar, where she learned how to lean over the pool table so that her breasts would nearly touch the green felt and men could see a nipple's edge. She took her time with each shot, and would imagine the fantasies running through the slack-jawed men, smiling and looking up at them as they pressed their pelvises against the pool table.

Gus left New Orleans for California, living in one of Los Angeles's sunlit valleys, where he smoked dope and dropped acid and hoped the Vietnam War would end before his luck ran out. Sabrina returned home to New Orleans, beautiful, independent, and smart, fourteen years old. As with Kinta, the apartment was a place to avoid, even flee. Sabrina spent most of her time with friends Uptown in some of the fine homes near Broadway and Tulane.

I wonder at Kinta's recollections, her library of memories. They seem so well-ordered and so fulsome, a repository to which she can reach to narrate her life. I listen to stories shimmering of youth and of abandonment and vulnerability: boyfriends, rouge and mascara, electric curlers and Aqua Net, how I learned to swim, the songs they danced to, where they made out, how they escaped the apartment late at night, men chasing them down the road, school and work, Mom's insults, the sailor, all the injuries alcoholics inflict on themselves and others. It's intoxicating to me, and it leads me to more documentary sources, interviews, walks down Second Avenue, to new questions about our family's past.

Even so, I can still scarcely tell a personal story from those two years following my return from Mississippi. The few memories I have seem more like frames cut from various parts of a motion picture, the remnants of some editorial process. There is no order to them until I bring them alongside the memories and stories of others, and the evidence I have collected that marks a time and a place. I try finding my life in the archive I have created. Sometimes meaning begins to emerge by the association of one fragment of information with another. A photograph with the date on its back suddenly becomes important. "Okay, so I was there" around so and so date, I tell myself. I develop questionnaires, asking relatives for dates, precise chronologies, whatever memories they have. I then triangulate the evidence, theirs and mine, and whatever else I've dug up. Kinta's job at Maison Blanche gave her an income and discounts. She could buy Christmas presents. I vaguely recollect her giving me a box of Hot Wheels, laying the plastic track along the floor beside a frosted Christmas tree. It must have been December 1969. Working backward, I can determine I was in New Orleans for 1968 and 1969 and part of 1967, in Mississippi from the summer of 1966 through the middle of 1967, then in New Orleans from 1965 to 1966. So this is my history, I tell myself.

From these swirling images it is hard to know what picture is my own. I had hoped that memory would somehow emerge in the gathered stories of siblings, by driving around places where I must have walked, or from the relics

I have recovered, for example a savings book from the Hancock Bank, account 83-042-2, recording an initial deposit of $7.50 on June 24, 1966, and a final deposit in February of the following year, ten days before my birthday, leaving a now unrecoverable balance of $19.

With information acquired from others, I thought I might begin remembering. My archive would invite memory. Soon I would string memories like beads along a thread, everything properly placed in time. But the memories I have stolen from others only add to the confusion and nausea of simultaneity, in which the past is both gone and hauntingly present in the gloaming of my conscious awareness. Time turns upon itself in ever tighter circles until I am left in a non-world of indeterminate tense, and the future, any future, runs away like a spring tide to an ever-vanishing horizon.

FOUR

---∿---

BOBBY

The school is still there, a rectangular citadel of brick on Jefferson Avenue. The children have gone home by the time I arrive. I walk empty corridors of polished yellow pine, glancing into classrooms festooned with learning; it's easy to imagine them filled with children fidgeting in their seats and teachers trying to follow their lesson plans amidst the boisterousness. At the end of a hallway a janitor sweeps away the memories of another day. An administrator describes to me the changes in the school since Katrina, how the school is now mostly African American and no longer named after John McDonogh, the eccentric and workaholic nineteenth-century slave owner whose estate decreed the creation of a public school system for the "education of all castes and races." When I ask about the records she tells me they are kept off-site. I'll need to go elsewhere to find traces of my earlier self.

It's not difficult reconstructing life at McDonogh No. 14 during the 1960s. There exist enough records to reconfigure a school day. I can insinuate myself into the evidence: oak desks for the pupils, and on the playground swings, monkey bars, merry-go-rounds, and zinc slides silvered smooth by children's backsides. There was the Morning Prayer to say, then the Pledge of Allegiance, and at some point in the school year lessons on the city's past: about when it had been founded long ago in the Spanish times, by men like Governor Francisco Luis Héctor, the thin-faced Baron de Carondelet, and of course about Lafayette's contribution to the American Revolution. There were the usual primers of American public school education, the Dick and Jane readers and math workbooks onto which children scrawled their answers. I would have walked past shotgun houses lined up along bare streets and solid middle-class homes cooled by the shade of magnolia trees and up the stairs to my classroom. Summer's spectral haze turned to winter's lucent skies beneath which Mardi Gras revelers danced in the streets. A few blocks away the Krewe of Thoth marched from Tchoupitoulas up Henry Clay and along Magazine Street and its tatterdemalion stores. Majorettes in sequined outfits and gold tassels tossed silver batons high into a cotton-candy sky. High school bands played songs like "Louie, Louie." A group of classmates celebrated the Apollo mission to the moon, dressing up as "Luna-tics." Between the bands and the horses there were the floats, and the men and women behind their masks and costumes, and beads of red and blue and silver, and gold doubloons tumbling through the winter light.

This much I know. In the afternoons Grandmother taught me French and math and Catholic prayers. She believed in rote, which is probably why it is her education I remember: multiplication tables, French nouns and verbs—*avoir*, *être*, *prendre*, *oublier*—and prayers, "Our Father who art in Heaven . . ." I recited my twos and fours and nines and elevens and repeated words that seemed to rise from her aged fingers moving across the rosary. I realize Grandmother was trying to keep an eight-year-old safe in the apartment, and that education offered a kind of stability. I could own what I could learn.

It's now possible to bring into some kind of association the few scattered recollections of boyhood with the bits and pieces of information I have gathered and tried ordering. Historical research confounds, however. The archive disciplines and provokes. A record tells you something but also shapes, in some senses seems to dictate by some magical force, the questions one brings to the past. A bit of data attains the exalted status of a document, insisting on its importance as if its author was right there beside you. It is an odd feeling of intimacy that releases me to endless wandering. I fantasize and long. Tenses intermingle as the time I believe separates what belongs to me, to me alone, and what I have gathered becomes indistinct. History places its own burdens on memory. The differences between what began inside my head, and what is out there in books and buildings, and the stories that fall from others' lips becomes less clear, less stable or assured. I begin questioning what exactly comprises my self, whether or not I am becoming a plagiarism.

I suspect that for most of the children, the school established the outer limit of their exploration, the boundary beyond which their parents conjured all sorts of urban danger. I roamed. Around 1968 I began taking the Magazine bus downtown to visit my mother during her lunch break. I stood on the corner waiting, watching men with their *Times-Picayune*s neatly folded under their arms or rolled up in a hand and women in knitted skirts in pastel shades of yellow, blue, and peach and sleeveless tops worn over heavy "I dreamed of" Maidenform bras. I imagine I liked sitting on the front row of four seats, where I could observe the bus driver's fingers click change away or hand the passengers gray, pink, and green transfer tickets. But I don't actually remember.

Mom worked in the cosmetics department at D. H. Holmes. Department stores offered women like my mother —women with few skills and little education, who at middle age had fallen on difficult times—steady if poorly paid employment. I can summon an image of her, an attractive woman in a long black twill pencil skirt and a white blouse, hair pulled back into a tight bun revealing a preternaturally high forehead, gold-encircled faux pearls pressed against her ears, lips painted red. She stands still and islanded in my mind's eye, illumined in the department store lights amid cosmetic mirrors and the twinkle of gold cylinders of lipstick and amber perfumes resting upon mirrored trays. We would often walk a few blocks to Acme or Felix's, have an oyster loaf or a plate of spaghetti and meatballs. She would drink beer, light a Salem, and unfold a slice of Wrigley's Spearmint gum from the open pack she kept in her purse.

After I had lunch with my mother I would usually stroll about the French Quarter, walking the length of Bourbon Street down to Esplanade and back again. I still wake to dreams of these ramblings, traces of the past riven by sexual fantasy. In those days, jazz and rock 'n' roll spilled into the street—drums and trumpets, electric guitars and saxophones, music that became intelligible then turned confusing and cacophonous as the sounds from one bar blended into another. Large signs advertised Al Hirt and Pete Fountain with his goatee, clarinet at his lips. I floated along with the crowds of men and women with plastic cups in their hands, stopping at show-bar entrances, where barkers announced the beginning of a new striptease or sex act. An ether, cool and wet, suffused with beer and smoke, escaped into the street. Inside the men seemed like automatons and the women too, with their alabaster skin, loose breasts, and dark nipples, and their pink slits when they lay on their backs and spread their legs.

In December 1968 or March 1969, Christmas or birthday, I got a bike, a Stingray with a banana seat and chrome fenders. Other kids pulled wheelies or threw their bikes into neat skids, leaving an arc of black tire across the pavement. Now I had my own. I learned how to stand on the pedals and race down the block, though I never managed to ride a wheelie more than a couple of feet.

The bike liberated me from my grandmother's supervision. In just ten minutes I could be in Audubon Park, or across Magazine Street and up on the levee, or past Tulane to the sno-ball stand on Plum Street, where I would sit on a bench eating tubs of spearmint-flavored shaved ice. I could look at the animals or sit among the great oaks, their boughs spread wide and low and welcoming, until a spring day's silvered light turned to bronze and in the distance the trees stood as silhouettes. I could bike to the back of the park, not far from the Mississippi River, where the WPA had created the highest spot in all of New Orleans, Monkey Hill, and join the other boys to race our bikes down its side as fast as we could, until tears started forming at the edges of our eyes and it was time to go back up again, for just one last ride.

I biked to the Prytania Theatre, where a classmate's parents worked selling tickets, or to Napoleon just down the street from Saint Stephen's, using money I had stolen from my mother's purse to watch a movie. I remember, as if I am still sitting there in the darkened hall, watching *They Shoot Horses, Don't They?* in the spring of 1969, just after I had turned nine. They are dancing for the $1,500 prize, and they are desperate and hungry and possessed like the people I had seen in the Quarter, groping each other, dancing round and round until the very end, when Jane Fonda passes the gun to her lover, who shoots her in the head.

—⁂—

Dad visited that August 1969, on his way to the Gulf Coast a week after Hurricane Camille. I don't know who he was going to help, Pat or Dorothy the Witch, but he came up the stairs and sat a while and brought me a brand new softball. I had not seen him for more than a year. I think I knew he had moved to California, though I had no idea where California was, just that it was distant and now he was here, sitting on a chair, and I was on the couch. He stayed no more than an hour, just to say hello and see how I was, and to give me the baseball, which felt smooth and dry in my hands. Then he put his fedora over his silvered hair and left.

In September I changed schools, my third in as many years. Saint Francis Assisi, on State Street, must have seemed far away. I had to cross Magazine Street, and walk a mile and a half. In the 1960s before school desegregation, there were only a few reasons for going to a parochial school: religion, wealth, or because you were in trouble. I was definitely in the last category, though no one ever said as much. There was, I guess, the discipline that was supposed to emerge by the mere fact of wearing a uniform. And there were the sisters and the priest and the church itself, with its neat rows of pews and the murals of Christ's tribulations I have since revisited, wondering what sense I made of them.

I had to wear a uniform of dark pants and shoes, and a white shirt. I have a vague recollection of standing in the playground watching the other kids, less a memory than a feeling of shame. There is a ball being thrown around, a basketball court, and childhood conversations going off in a million directions. It is the clothes I remember. Everyone else seemed cleaner, properly put together, even as shirts starting loosening their way from pants and belts.

Although I had been baptized and my grandmother and I did the rosary, I was less than an ideal Catholic. I hadn't taken my Holy Communion. Divorce had effectively exiled my mother from the Roman Catholic Church, which might have offered her some solace. So the sisters insisted on my confession and communion. The latter was easy enough. You simply followed the others from the pew to the front of the church, kneeled, and repeated whatever the person next to you said, or simply mumbled something incoherently. Then you walked back, kneeled one last time while making the cross, and scooted your bottom along the pew.

Confession was a wholly different matter, private and, I imagine, menacing. What went on in there, what could happen behind the pulled curtain? The sisters must have given me a booklet to read, instructions of exactly what to say, the whole spiel beginning with "Forgive me Father," except that by the appointed day I hadn't memorized a single word. I think I folded the booklet into my pants pocket, and once inside the booth immediately pulled it out, running through the words. I lied at my first confession, bullshitted the priest.

It is guilt that has helped me remember that image, just as it was a child's feeling of disgrace for wearing tattered clothes that has settled a memory fragment within me. There is no way of confirming either recollection, except for the sensation I still feel when I drive up to the church and school. It is the feeling of slight unsettledness, knowing the past is never far behind, the unsteady stir of memory rising, like my recollection of the time my mother came to the Webelos scouts celebration. I am standing with the other boys, fidgeting before the auditorium lights, and there she is, near the front, and she is very drunk. A sister is there as well, unsure what to do in this public spectacle. Mom says something, loudly, belligerently, and I think she slips off the folding chair, though I can't remember exactly what happened. There are really only two images bequeathed by that experience: standing there on the stage and seeing my mother slovenly drunk, and then afterward coaxing her into a taxi.

—⟆⟆⟆—

There was a large fig tree in the backyard, with broad sticky green leaves and purpled fruit that I picked in summer for Grandmother. She peeled the fruit for morning breakfast, canning the rest so in the winter months she would have fig preserves with her toast and the coffee she brewed in the tin drip pot. I wonder if this simple ritual reminded her of a life-time ago, when her sister Mae made preserves from the fruit trees in her backyard.

The tree is gone. "Toppled right over on that fence there," Mr. G. tells me. His arm points accusingly as if the tree slighted him, fell with the clear intention of costing him money. "Hurricane Andrew. Had to pay some guys to cut it all up and haul it to the curb."

I can imagine climbing through the tree, my limbs as thin as a gibbon's. I am plucking ripened figs and spying on my neighbor Bobby. He is working the mower along a nar-row rectangle of grass, the spinning reel of scimitar blades rasping like a pair of shears with each push down the yard. Afterward he sharpens the blades with a long file, oils the machine before returning it to the back shed. Or he is clean-ing a birdbath next to a statue of the Virgin Mary in white and powder blue. Or he is in a lounge chair with a cup of smoky chicory coffee leafing through the Sunday *Times-Picayune*. What is constant is a desire. I hope he will be there, and that he will see me watching him.

Through the summer months Bobby wore loose cotton shirts, often in plaids of blue and light green, or no shirt at all. He was in his twenties then, too old for the draft, a handsome enough man though already balding, with shoul-ders nearly as wide as he was tall and a smile to match. He had a predilection for very tight bathing suits, Speedos of varying colors, tight enough to reveal an arc of flesh.

Bobby had the entire upstairs floor to himself, and a separate entrance. It was a long house reaching deep into the city block, as if someone had decided to build one shotgun on top of another, with a small porch in front that ran around one side and the standard kitchen at the very back. Bobby's parents were Italian immigrants. They lived below with his sister Shirley, who was "not right in the head" and went in and out of mental institutions. Religious tapestries covered 1960s wood-paneled walls, chocolate browns and pelagic blues and brumous colors that sapped the light that had fought its way through the windows and heavy curtains–a Christ child with parents, scenes of crucifixion and judgment, the Last Supper with Christ's open hands, a gaunt Jesus with dreamy downcast eyes—images that seemed more like apparitions when they undulated in the breeze from the window fan.

I don't remember when we first met, the first time I lowered myself over the fence and fell into his world. Research offers an approximate date: 1968, the year I returned from Mississippi. I was eight. Bobby gave me a few cents to buy candy from the corner store on Magazine Street: Red Hots, Bazooka bubble gum secure in waxy wrappers, Lemonheads in their neat little boxes. If I had enough money I might get a moon pie, tearing the cellophane wrapper with my teeth, or ask for a large dill pickle from the jar up on the counter. I might wander into Woolworth's, where Bobby worked as a clerk, if I was downtown visiting my mother for lunch and tired of walking up and down Bourbon Street listening to jazz and looking at women. Bobby might buy me some candy or a model plane of gray plastic, an F-105 or an F-4, plus a small metal tube of glue.

I like to think Bobby enjoyed walking the few blocks to the bus stop smoking from a pack of Lucky Strikes and returning home in the late afternoon or early evening; it is a kind of gesture of thanks projected backward. The Plasticine world and fluorescent lights of Woolworth's suited him. The work was easy and the money adequate as long as he lived upstairs at home. He could walk along Canal Street window shopping, basking in America's exuberance and its neon abandonments, and on the way back home with the bus window slid open he could dream of weekend parties and a hand running across a chest and down a leg.

I suppose he was attracted to some of the young men he met downtown. Many had left their family's disavowal and the small town pettiness and intolerance of the American South, arriving in New Orleans hungry and lonesome and hopeful. It was difficult being homosexual anywhere in America during the 1960s, but especially in the South with its tradition of violence against people defined as somehow deviant. New Orleans was something of an exception, as always, with a substantial gay population. Police looked past bars known to be frequented by gay men, especially when the owners paid them off. And no other American city cultivated a whole season of saturnalian delight when cross-dressing was not only permitted but applauded, though after the fun was had people were expected to return to heterosexual lives, or at least keep their predilections private.

Bobby kept a coterie of boys and young men; I was the youngest, more a charge than a future lover. They would arrive for weekend parties, hair greased back, standing at the window or on the porch with a longneck in their hand and a cigarette between their fingers, listening to music like Frankie Valli. Bobby sat recumbent at one end of the couch in his bathing suit, an arm draped across its back, happy and sated.

I think Bobby taught me how to mend socks, and how to tie the laces to my high-top Converse All Stars. The trick was in the preparation, making sure the laces crossed over tightly, holding the loops with your thumb and middle finger. I am in his backyard, sitting on the edge of a chair so that my foot can touch the ground. Bobby is kneeling, holding the laces in his hand, and he is apologizing. They are among the only words I remember from my New Orleans childhood, as clear as yesterday.

"God made me this way."

I knew what he meant, then as I do today.

My mother warned me away from him; in fact, there was consensus among family members that I was at risk for sexual abuse, or worse that I was becoming a homosexual. Ultimately Bobby became the reason I left New Orleans again. She would say something like "Stay away from that homosexual." Or "fag." But I continued visiting Bobby. By the spring of 1969 I had become a part of his life. I must have loved him. From the scraps of evidence I have collected, including photographs Bobby himself kept until 2007, I discover not one but two identical school photographs I had given him, one in color, the other black-and-white. I have no memory of this transaction, of walking over to his house, of placing the photographs in his hand, or of what he might have said, nor do I know his thoughts when he wrote my name on their backs and dated one, 1969.

Photographic evidence tells me that in May 1969 I attended two birthday parties at Bobby's house, as if I was family. The first is for his sister Shirley, who in another picture looks to be in her twenties. It is Monday evening. The weather has turned warm. Shirley is in a white dress and white shoes, leaning over the table to blow out a single candle on a layered cake thick with white icing. Everyone is looking at the cake, except for me. I am staring at the camera in my bare feet and stained New Orleans Saints shirt, looking straight at Bobby. The flash has turned both eyes into silvered dots. I am leaning back on my right foot with my fists raised like a boxer. The pose is less aggressive than flirtatious or precocious, as if I was acknowledging our special relationship. A week later, I am at "Jr.'s Birthday Party." In this photograph I am eating yellow cake with white icing. There are four other children there, all smiling around the cake and tubs of ice cream, an iconic American picture of celebration and family life.

In the late 1980s I briefly visited Bobby. I had not seen him for two decades. It's difficult understanding why I stepped up to his door that New Orleans summer day. Perhaps it was because I was still looking for a father now that my own had recently died, that somehow I was reenacting a past abandonment and longing, when my father had sent me away from Mississippi and Bobby had looked after me, offering clothes and toys and sweets and the simplest of skills, how to tie one's laces. Or that I hoped the visit would somehow allow memory to speak, his and my own. I could discover what had transpired between us, and what my life had been like in his eyes. I could ask questions with the security of an adult, a married man with a steady job and thoughts of having a family. Why had I wandered into his life? Why had he taken me into his fold? Why the apology?

Bobby welcomed me inside. We went upstairs where I had gone so many times as a child. I was nervous, frightened by the questions I wanted to ask him, hoping something would come from my seeing him again after so long a time. There were two other men, relatives he told me. One, about Bobby's age with tattoos painted across his arms, was doing macramé. He seemed to spend his time on ships. Bobby and I went looking for the other man, a cousin. When Bobby opened a door, he was lying on the bed masturbating. We walked away as if we hadn't seen anything.

We spoke for a while. I took a few photographs of the old pictures he showed me, plus a few others that seemed meaningful: the house, tapestries, six of his young darlings posing on the staircase. None of the questions that swirled within me seemed possible to ask. I told him about my career, offered to help his cousin with his GED, steered the conversation away from what had brought me there. I could not tell him of the concerns my family had had that I would become like him—a deviant, a pervert. Bobby was grooming me to be his lover, they worried. I thought we could laugh together at their silliness, their prejudice. Yet I could not bring myself to ask him about the apology that persists and disquiets, producing within me a strange sensation that something may have happened a lifetime ago.

—◊◊◊—

Not knowing has brought me to studies of trauma and memory, and I soon realize that my desires, all this neurotic fretting about wanting to know what happened, are more than a consequence of my and my family's interest in my sexual past. They are part of a cultural and historical moment. Trauma and memory have become international obsessions, especially when sex is involved. Never before has there been such an urge to speak of the unspeakable, to represent what seemed beyond comprehension, and to believe that trauma defines who we are, especially when we are young. The child, we like to say, makes the man.

We surround ourselves with traumatic pasts. Countless television shows parade victims and perpetrators, most often as vulgar sensationalism but also sometimes in the guise of therapy. Modern technology has transformed the exceptional into the everyday. In our pockets and purses we can instantaneously access a shooting or a tsunami, watch a murder or a battle, or tweet about a violent death, whether any of these things happened next door or ten thousand miles away. These events become part of our memories in ways that are dramatically different from reading a newspaper story about a tragedy that took place in a faraway land. At the same time, we can share our inner lives with the multitudes, in real time.

The medical manipulation of traumatic memory has become an exciting area of scientific research, and for some quite troubling. Scientists have identified a gene and its variants that is tied to susceptibility to PTSD. This has helped lead to identifying biochemical markers and developing drugs that will interrupt the making of memory, specifically the consolidation of long-term memory. In one experiment scientists were able to produce PTSD in mice; injecting the mice with a drug before or after a traumatic event seemed to prevent PTSD symptoms. Propranolol, a very widely used drug, has been found to alter the process of memory consolidation. Now that we have identified various pathways in the brain tied to memory, we can administer drugs that assist remembering and forgetting. Rape victims or soldiers in battle will soon be able to take a pill or an injection that will prevent PTSD or mitigate psychic turmoil by changing the way they recall the traumatic event. These treatments, it is promised, will reduce medical costs and mental anguish —and for soldiers permit their speedy return to war.

A number of neuroethicists are alarmed by these developments. In 2003 the President's Council on Bioethics issued a statement of concern around the potential biomedical manipulation of memory, worrying that new drug therapies might "alter our sense of self." Others are less apprehensive. The medications, after all, would prevent the formation of a single memory, not the constellation of memories that forms our identity. And, anyway, why would someone want to remember something that produces suffering? As the controversies continue, what is clear is that we live in a radically new world of trauma, memory, and forgetting, a world that is unfolding globally and, quite literally, at the level of individual cells.

Childhood and memory remain especially fraught subjects, debated among academics, fought over in courthouses, and discussed endlessly around the dining table. Memory is central to how we think of ourselves, but it can be highly inaccurate. Memory can be insistent or fleeting. Memory erodes; forgetting is an important, even vital, part of life. We want to insist on our memory's truthfulness even as we are quick to point out the errors of others. Our desires and fears shape how and what we remember. Because our minds seem to work through a complex web of associations—as in the reveries cascading from the simple act of Proust eating a madeleine with a cup of tea—the meaning we assign to the past is subject to change. We confuse our feeling for what actually happened. The feeling becomes the fact as our emotions form a narrative around a scant trace of the past.

The question of childhood memory is particularly challenging. How reliable are childhood memories if childrens' brain and memory systems are undergoing such profound development? Can adults recover repressed memories such as incest or sexual abuse by caregivers? How does the traumatic event endure within the life of a person, in their unconsciousness and in their daily lives, even in succeeding generations? Does a dream or a fantasy contain within it something real?

These rather academic-sounding questions reverberate in our public culture. In the 1990s, controversies raged over instances of alleged sexual abuse in nursery schools. Criminal trials ensued, including jail sentences for some of the accused. Across the country, teenagers and adults began accusing others—usually a parent—of committing some unspeakable act when they were children. Their recovered memories of molestation explained who they had become— their sadness, relationships, even careers. Estrangement and divorce often followed. Fathers ended up in prison. Entire families broke apart. Many of these accusations, it turned out, were entirely baseless.

This history forms part of a far wider set of discussions and fixations on trauma and memory, including a widespread fascination with locating the mind in the brain's neural networks. There are endless commercials about memory loss, drugs to be prescribed, vitamins to consume, mental games to maintain our selves. There is a multi-billion-dollar industry just about memory. Alzheimer's and PTSD have become the metaphors of our time. The politics of memory have attracted enormous attention and debate. Some of these conversations have unfolded within our universities, in the commitment of research funds, the creation of memory and trauma "studies" programs, and in exchanges between literary scholars, philosophers, legal theorists, ethicists, and neuroscientists.

Troubling pasts also have become the work of governments. In South Africa following the end of apartheid, the African National Congress created a national experiment that aimed to recover memory and make public the secret world of state terror. Men described how they tortured and killed others. People offered various histories, some personal, some of entire communities, as though past events remained persistently present and wounding. Men and women spoke of horrific violence, of rape and murder and disappearances. The goal wasn't justice, the use of historical knowledge to rectify past wrongs or to allocate punishment. The goal was truth—raw, unvarnished, visceral truth. Memories made public and transformed into redemptive histories would explain what happened to a son, daughter, husband, wife, friend, or lover, or communicate to the world—and most immediately to the perpetrator—the sufferings of those who survived and were left behind to pick up the pieces.

In the United States trauma and memory have become a vast industry involving pharmaceutical companies, therapists, and the major branches of government. The past has become uniquely troublesome. PTSD emerged as an "official" disorder in 1980 largely because of the vigorous activism of Vietnam War veterans and doctors, though it had its roots in earlier work on "shell shock" and "combat fatigue" during the First and Second World Wars. PTSD is commonly invoked across a wide spectrum, from war veterans and sexual abuse victims to bystanders witnessing a car crash. Suddenly everyone seems traumatized. According to conventional wisdom, the adult PTSD sufferer can't forget. The traumatic experience keeps returning, what scientists call "involuntary memory," while autobiographical memory seems shattered. They become their trauma.

Alzheimer's exploded into public attention at roughly the same time as PTSD. Both afflictions involve the same memory systems within the brain. The limbic system is the first area of the brain destroyed by Alzheimer's, particularly the amygdalae and the hippocampi. Alzheimer's steals the past, in the end quite literally annihilating the self. The victims forget everything. They forget their loved ones, their histories, ultimately their very selves.

PTSD and Alzheimer's occupy the twin poles in our national conversation on trauma and memory: Either we can't forget, or we can't remember. Our preoccupation with trauma and memory helps explain the emergence of memoir as our defining literary genre. We all ask "Who am I?" and answer with a story, or more precisely a history. *This is who I am now because that was me then*, we say—a child or teenager, someone's son or daughter. We point to people and events that shaped who we have become. Memoirs are histories of the self, stories of how people became who they are. We are drawn to memoir because we live in terror of forgetting, of losing our private thoughts and memories of others, ultimately of forgetting who we are. Today we live without the networks of extended kin and community that once defined who we are and preserved and transmitted history to succeeding generations. In the end, the only thing we own is the past that resides inside us.

In memoir, memory seems resolute, somehow able to withstand the erosions of time. In many works it's as if the authors' childhoods are before them still, or as if they knew from a very young age that they would be writers and that one day they would write about themselves using all the notes and impressions gathered over a lifetime. This clarity exists even in those memoirs in which there is much sadness, the loss of loved ones, of love itself, the brutalizing of adults, and still darker tales of sexual abuse at the hands of relatives or even the writer's parents. The authors seem so certain, describing what they were like then, how they felt at a given moment many years before, as if they could bring a mirror to their past and declare, "That's the way it was." This is in part literary convention, the omniscient narrator's voice offering the reader a kind of authority or mastery, but there is also the peculiarly American tradition of both knowing the minutiae of one's past and making it no matter the odds, the usual narrative of struggle and redemption: "This was me then. And this is what I've made of myself."

The historian in me says something's amiss in this claim that the past is so readily accessible and transmissible through writing. Few people are so precocious, or so lucky. Writing may be the antidote to memory, but it can never be its substitute. All we can do is offer images of a past that in some basic way remains absent. Memory—and that is what memoir is after all, memory brought to language—can't possibly arrive so readily, so clearly and truthfully. There are different kinds of memory that involves various regions of the brain, a symphonic dance of synapses, biochemicals, and external stimuli shaping what we see, do, say, and write. Because the brain is ceaselessly changing, the issue is not simply one of processing the external world. That world is also inside us.

Memory is as capricious as our desire to get the story right. We spend our lives comparing and contrasting experiences, trying to somehow locate the past in its full context, so we can produce within ourselves *the* history, not *a* history, a recounting of a probable past. We weigh our lives, much as historians weigh the past. So we tell stories, and tell them again and again, knowing that we live in their recollection.

Perhaps I am just jealous. Memory has yet to emerge from all the interviews, archival research, poring over records, walking up and down streets, sitting in bars, reading scientific papers on the childhood brain, not to mention all the decades of therapy. History is memory's impoverished replacement. I thought that if I worked hard enough memories might begin revealing themselves from the recesses of my brain like some ancient relic from the sea. Somehow I would be able to put everything together, create a decent enough picture of the past so that I could say, "Yes, that was me back then. This happened, and it was horrible, and, yes, I suffered." I could connect memory to feelings of guilt and shame, make things right by the simple acknowledgement that I was only a child and everything was beyond my control. I was wrong.

—⚬⚬⚬—

I could not answer what had happened between me and Bobby; at least I could not do so alone. There were just too many blank spaces, too many absences. And the problem with sexuality is that because it is private, no one else really knows.

It was a bit more than a year after Katrina. I was well into the research. I had a stack of Xeroxed papers, hours of interviews, and books of notes I had taken. I had spent months reading into the literatures on memory and trauma. And I was in the middle of psychoanalysis. But nothing was emerging, no memories of childhood I could now work through in the safety of a therapist's office. I became convinced that the absence of memory meant the presence of something terrible. I was driving myself crazy.

So in 2006 I wrote a letter. It was a simple text, a plain request for information shorn of accusations or histrionics. I had some questions, and I needed answers. I thanked him for teaching me how to tie my shoelaces.

Bobby never replied.

Months passed, then nearly a year.

I grew suspicious, anxious, convinced his silence meant I had been molested. I would come home from meetings with my analyst exhausted. I had dreams of a green room, and that something had happened there, repeatedly. In another dream I am looking from my back steps onto Bobby's house. A pack of wolves suddenly appears and chases me through the apartment. Once, after a session, I vomited outside, near the car. The analyst spoke of bodily memory. Hair began falling out in clumps. She assured me that victims of sexual abuse rarely become abusers themselves. But I worried as a father. I spoke to my wife, convinced that I had suffered abuse, from a man, from Bobby, based not on a recollection but on a sense confirmed—and encouraged—by a therapist.

Toward the end of 2006 as I was driving on a highway outside Atlanta, I found myself dialing the operator. I wrote down the number, dialed it, waited.

"Hello."

"Umm," I stuttered for a moment, thinking I should hang up.

"Umm, hi, is this Bobby B.?"

"Yes. Who is this?"

"Hi Bobby. It's Clifton, Yvonne's son. We lived on Chestnut Street, around the corner."

I had last spoken to Bobby more than a decade ago, when I had traveled down to New Orleans for my father's funeral in the summer of 1991. I wasn't sure he would remember me. But Bobby did remember, and though many years had passed his voice seemed familiar.

I tried explaining to him that I was researching a book, a book about New Orleans, a book about my life, including the two of us. I said I had sent him a letter, and that I was now following up, phoning. And then I had to ask the question, except in my mind all sorts of variations flashed before me. "Did you touch me? Did you sexually abuse me, rape me, fondle me? Are you a pedophile?" But I found myself not wanting to accuse him. Indeed, I felt no anger at all. I simply wanted to know, to have some information, finally. So I found myself reaching toward some kind of neutral language.

"Uh, Bobby," I began, "did something inappropriate happen between us?"

He knew exactly what I meant. And he answered clearly.

"No."

My question now felt like a betrayal. I wanted to take everything back, start again by thanking Bobby for his kindness.

"I don't know what your problems are," he began saying, as if I were fucked up, until I interrupted and began arguing, explaining I was quite fine, with a lovely family and great job.

It was a lie. I wasn't fine.

I thanked Bobby, said good-bye.

About a month later an envelope arrived in the mail. Bobby sent me five photographs; the two I had given him as a child, three from the parties at his home, and a long letter. I held in my hands all the documentary evidence that once connected the two of us. "I lied about [not] receiving your letter," he began (I have reproduced the letter as it appears):

> To be truthful I didn't know what to say to you. You see there has been a lot of things going on. The old house was torned down and a new one builted in its place. There are a lot of things I would like to say to you . . . I tried to be a friend to you, but I think you either was shy are someone was miss treating you. I do know one thing. You turned out to be the best and I am proud that I know you.

Bobby was glad I was so successful, so smart, and that he had in some small way played a part in my life. And he was grateful I remembered him. But the letter actually says very little, and virtually nothing about what I want to know, his recollections of my childhood, our relationship, what happened.

When I first opened the letter, I had both hoped and dreaded that he would continue our conversation. Bobby would write *No, I never touched you, ever*, and then admonish me for the suspicion. *You needed a father. I tried to take care of you the best I could. I offered you protection, love. Don't you remember me teaching you how to tie your sneakers? I have kept photographs of you for nearly four decades now, and you come to me with this accusation . . .*

Or he would explain to me what transpired in his room all those years ago. I would finally have an explanation, however disturbing. All the riddles of my past would somehow be resolved. There would be an origin to my suffering, a reason for my inability to remember.

And I could then write back, explaining that my search was not for justice, and certainly not retribution, but for truth, for some knowledge of the past. I could tell him what he did was terrible, basely wrong, and that I knew how love and abuse can go hand in hand. I could offer a modicum of forgiveness for our kindred shame.

"What ever you write in your book will be tops so don't get discouraged and I know it shall turn out real good. It sure is good to know that you have found happiness altho as a child I don't think so but God has a way of rewarding those who wait on him. I do a lot of praying," Bobby wrote toward the end of his letter, "and that is the only thing that keeps me going."

Bobby's letter is a good-bye of sorts. He has, in a sense, gotten rid of me. He has returned the photographs I once brought to him as a present, as a token of my affection for the man who taught me how to tie my laces. There is now no record of me in his house, nothing he can look at on a boring winter's day, thumbing through old photographs in the final years of his life. My research, all research, bears costs; I have been expunged, disavowed, and I have hurt people in my past, more than once. Bobby doesn't want to know me, wants indeed to forget me, wants no trace of our past.

Bobby promised to answer any questions I might have, and that he would do his best to respond promptly. He wrote that he was a Catholic, though not a good one, and that he had been collecting Mardi Gras beads for me whenever I visited, as if I were still nine and he wanted to offer a simple gift of New Orleans.

I did visit the neighborhood again, walking around taking photographs. I passed Bobby's new house a few times, thinking I should knock on the door. I saw a man upstairs on an exercise machine, but it wasn't him. Then as I was about to get into my car I saw him. Bobby saw me too, from a distance, and for an instant I think he recognized me. I got in my car and drove away.

FIVE

LEAVING NEW ORLEANS

MARIE HANDS ME THE CARDBOARD BOX.

"Take them. I'm not sure what's all in it, just pictures I suppose, stuff Dad had."

It's a small box she must have brought home from the hospital where she works as a nurse, just big enough for fifty pairs of pre-powdered surgical gloves.

Bill senses her discomfort. He's a kind, if imposing man, 6'4" and 250 pounds, with a linebacker's neck. He and Marie have been together since they met at a high school dance in the late 1950s. Marie was seventeen, Bill a year younger, children except for their bodies. They made out, and made love, wherever and whenever they could. Marie soon became pregnant. They tried eloping, driving to Alabama where the magistrate told them they were too young and sent them back home to New Orleans.

They married in the end, against the protests of Bill's parents, who ran a successful bakery delivering bread around the city. Bill was the favored only son, expected to carry on the family business, marry well, and live in a well-appointed Uptown house. They were certain Bill was ruining his life, or more precisely that my sister and our dissolute family were. My father congratulated Marie for snagging a man from a good family by getting pregnant.

"Too much dwelling on the past," Bill says. He would rather be out on the water fishing or walking on the beach, doing anything but revisiting those difficult years when they had one child in diapers and more on the way but spent their weekend evenings scouring the city, plucking my mother from bars. They would try to get her into the car without the usual spill of acid words, and back in the apartment Marie would help Mom into bed and lay a cool compress across her head.

Bill thinks I'm wasting my time. He watches over Marie, who is just over five feet tall and bent over as if tugged to the ground by her life's travails, thin as rails and as fragile. Whereas Kinta's reserve seems a kind of self-protection, and Sabrina revels in a hippie's exuberance, Marie harbors a terrific guilt over not having saved everyone. I wonder if part of it began with the burdens the court created in May 1966 when the judge, in response to my father's petition, told Marie and her husband that they were responsible for the minor children. In Marie's and in everyone else's recollection she had guardianship, not permanent custody. In fact Marie was supposed to have taken us in, and she was to receive child support payments from Dad. None of this happened, and it has not been possible to discern the conversations that unfolded that May and why we didn't move in with my sister, who had just turned twenty-one and had two children of her own. But the judge must

have created a terrible weight on my sister, a set of obligations she could not possibly have met.

"You can't do nothing 'bout it," Bill tells us, standing
by the sink, rinsing the dishes from a meal of Shrimp Creole
we spent all day making and about an hour devouring.

"Maybe some records too," Marie adds, pretending
she hasn't heard Bill's warnings. "I got them from one of
Letha's daughters. You know, when they cleared things out,
after Dad . . . She sent it all." Dad's third wife, Letha, had
died. One of her daughters, Sherry, ended up with the stuff.
Sherry separated a lifetime's records, keeping everything
from before Dad's marriage to Letha. The rest headed to
Marie in Florida.

"Go on, take them. Keep them as long as you like.
Really. They're yours."

Back in Atlanta, I start rummaging through the box.
A few photographs are in plastic cellulite frames, four-by-
eights of high school graduations, weddings, studio portraits
taken against a gray-blue backcloth, children buttoned up
in tight collars looking like they want to be out on the playground with their friends. The rest are from an old album,
the kind with a spring binder that allows you to add new
pages as children marry and have babies, vacations, gather-
ings, family accomplishments.

I begin by trying to be a scholar, as if I were assessing
a box of records brought to my desk from some musty
archive: number of documents, provenance, how the material is ordered, dates, identification of individual records.
Historical research typically unfolds this way, or at least
that's what we historians usually tell ourselves. Organize,
categorize, these are the basic rules of archival research, the

due diligence that precedes interpretation. During the nine-teenth-century professionalization of the academy, historians made claims to being scientists by employing increasingly complex forensic methods. Scholars believed that these procedures of evaluating evidence would allow them to recreate a past world that they could evaluate objectively. They could thereby produce a true history in contrast to more literary approaches that seemed rife with bias and perilously close to that most subjective form, the novel.

I resist the temptation to begin ascribing intention while these arcane debates swirl within my head. But very quickly, too quickly, I start looking for records of me, and from there all sorts of wonderings tumble forth—what if my parents hadn't divorced, and we had money, and I had been a wanted child. Then an entire story unfolds about a middle-class white family and picnics in Audubon Park eating fried chicken before this fleeting wistfulness turns to disaffection. I become jealous upon discovering there are only a couple of photographs of me compared to dozens of my siblings, nieces and nephews. Jealousy turns to anger and feelings of disavowal, as if these photographs stand accused and are the reason I didn't once rest on the mantle in his California home.

The manufacturer had sprayed a mild adhesive on the pasteboard. I imagine Letha looking over a pile of newly developed photographs, deciding which ones to throw out and which ones to affix to the page, bringing a plastic sheet over top, and then showing my father her dutiful work. The plastic sheet was supposed to seal the images in place, to protect them as my father and Letha, and now unexpectedly his son, leafed through pasteboard lives.

There must have been a design flaw, or time's passing has weakened the bonds that once held the images in their intended order. Many of the photographs have started coming unglued, slipping free from the page and sliding into the jumble of images at the bottom of the box. A few more lie scattered each time I go looking for something that might awaken memory and help me figure out what happened all those years ago. It is as if the impersonal forces of entropy defeat any attempt to restore order. I sort through the hodgepodge, but with each effort I seem to be less successful. The past deteriorates steadily, inevitably.

Sometimes I think about reassembling the album differently. I would loosen every photograph from its page, then put the whole thing together again in a new album. I would add photographs; I could even make them look old. The orthodox historian in me says this would violate basic rules of evidence. This documentary of me would be a plagiarism, a mere fiction, though I also wonder just how much our fantasy lives shape the histories we imagine and write, our desires and fears and hatreds reverberating in the characters and dramas we summon from the past.

Many of the photographs are of Dad and Letha. In the background brown scrub foothills lie beneath Southern California's unfettered light. The two of them are invariably smiling, celebrating a special event, enjoying each other's company, in love even, whereas the Dad of the 1950s photographs seems saddled and dour. I've not been able to figure out why he moved all the way to California. Perhaps it was Dorothy the Witch's idea. *Let's get away from everyone*, she might have told him, *away from New Orleans and the heat, away from Yvonne and your wretched family.* Any-

thing was better than hidebound Mississippi, and returning to New Orleans was out of the question. California
was about as far away as one could go without leaving the
continental United States. Maybe he thought there might
be safety in distance, or just wanted to make a new start.
Dad knew about the huge growth in the defense industry
from working in an air-conditioned trailer at NASA.
Whatever the reason, the move was part of a massive
westward migration; by 1970 one out of every ten Americans lived in California.

In 1968, Highway 10 whisked Dad, my sister Sabrina,
and Dorothy the Witch right across the country. They stayed
in one cheap motel after another until they turned off the
freeway and found a shabby two-bedroom apartment in
Burbank just off the main road. Dad was in his mid-fifties,
with just a couple of hundred dollars in the bank, few qualifications, and a track record of broken employment. But he
had a routine for jump-starting his life, dyeing his hair black
and slicking it back with Brylcreem until it looked like polished obsidian. Youth had a perpetual edge, he figured. It
was important to appear young, to look like a married, mid-
career man far from the desolation of age, ready to make
good in a land where everything seemed perpetually new.
Dad would buy the Sunday *Los Angeles Times* and spend
the morning drinking Folgers Coffee and penciling circles in
the employment section. He found a job at Chevy Chase
Staff & Stone as a bookkeeper. The housing boom in Orange
County and San Bernardino Valley created a huge demand
for landscape products—pebbles for rock gardens, stone
squares leading through closely cropped lawns, cement
bird feeders, the suburban American idyll. The company

needed someone to keep track of all the bills of sale, someone like Dad who was good with a Remington Rand
adding machine.

Mom kept calling him, complaining about child support checks that hadn't arrived, unpaid bills, utilities about
to be disconnected, errant children, every hurt she had endured
during a seventeen-year marriage. For three decades after
the divorce she kept all the bills in his name, as if she were
still claiming Dad as her own or expecting him to provide
for her, to be the man she had so wanted and felt she deserved. "Get the money from him, my husband. He's in California," she would tell the collectors, before giving them
my father's phone number and address.

At least once a year she wrote Dad a letter telling him
what a bastard he was for leaving her with a bunch of kids,
moving her all over the South, never owning much of anything, for every misfortune and ruin.

She'd call him every few months, on the weekend when
she was lonely and desperate and unpaid bills had come due.
It became a kind of ritual, requiring beer, a fifth, and a pack
of Salem mentholated cigarettes. Mom would dial the rotary
phone when she had gotten very drunk, not quite ready to pass
out but well into slurring her words, somewhere between
unrestrained spleenful anger and outright incomprehensibility.

Howard! I imagine it would begin with an accusatory
yell that also sounded as if she was nauseated.

*Howard, Howard it's the fifth and . . . yousonsabitch
Howard, the goddamn check isn't here. They're gonna cut
off the goddamn phone any day, any goddamn day now.
And how am I goin' feed these kids. That son of yours, he
got bronchitis again. You gotta pay the doctor.*

Dad would say something, probably that the check was in the mail. There would be an extra pause between what my father had to say and my mother's response, as if the booze had slowed down her neural connections so that the words didn't quite arrive in order, and then you could see her about to say something, something real angry, downright reptilian.

Look what you've done, Howard! You goddamn coward, leaving me with these kids, moving us all over the place. My family lived on Esplanade, don't you forget that, and I'm livin' poor as dirt. And you can't even pay your goddamn bills, your obligashuns. You know what I mean awright. Don't you hang up the phone on me, you fucker. Howard! I want that check. You owe me. You owe me everything. Howard!

She would slam the phone down, yell for a beer, take a swig, light up another cigarette. An hour later she couldn't get her index finger into the rotary phone or read the number.

Clifton! Clifton! Your goddamn no-good father, Clifton, come dial this goddamn number, can't pay his bills.

She would be sitting on the bed in a slip, unable to stand, her eyes glassy and cheeks mascara-stained. Her head would swing around toward me like some menaced, cornered animal.

Dial this number. You know who your father is? I'll tell ya, he's no good. He didn't want ya, didn't want any of ya. All he cares about is himself. And don't ya forget, ever, he didn't give a damn about any of ya.

I dial my father's number, then hide in my closet room. Lying in bed with a pillow around my ears, I hear my mother

yelling. "I bet she's not as good as me, is she Howard," she'd
tell him; half biting, half hoping my father would answer
the way she wanted him to, with something redeemable
from their estranged past.

I suppose my mother must have called Dad and insisted
he take me again, or one of my sisters told him to do something. She has no memory of this, nor can she provide any
detailed reason for why I had to leave New Orleans except
that I was "gettin' in trouble." There was never any conversation with me about what was happening, what I thought
I wanted. No *Don't ya think it would be nice to go and visit
your father for a while, out in sunny California* kind of plat-
itude, the sort of insouciant leading question that always
means big things, and which generally sends waves of emotion across the faces of boys and girls who spend their childhoods shuttling between parents.

My sisters don't say much when I ask them, though
they and others volunteer information on my numerous
youthful indiscretions that seem humorous with the passage
of time, especially in light of the apparent evidence that I've
"made it." Their eyes turn blank when I try pressing them
for the exact reasons I moved to California, as if they were
looking right through me and had returned to when their
lives were elsewhere, to that indeterminate time between
Barbie Doll dreams and adulthood when they survived by
staying away from the apartment as much as possible. I
realize there is something they won't say. Or can't.

"Truthfully, I just don't know," one answers obliquely.
"What do you think?"

I want to argue back. Why do my sisters begin so many
of their sentences with "truthfully"? It's a Southern collo-

quialism, but part of me wants to rail that it's a Southern inability to tell the truth.

"Gettin' in trouble" seems accurate, though. It was 1969 and 1970. I was ten and in fourth grade, skipping school, forging (with spectacular ineptitude) my mother's signature on my report card. *My Mom don't write so well*, I guess I told my teacher, cocking my head a little to one side. At home, I set the downstairs utility room on fire. Smoke started coming through the floorboards. Grandmother was just about to call the fire department when Kinta got home from work. My brother taught me how to siphon gas and hotwire cars, skills that were particularly useless since I didn't know how to drive. There were other more practical skills like stuffing vending machines that would prove lucrative. Sometimes I'd return at the end of the day with a pocket full of dimes and quarters, enough for a meal at Domilise's. I had started sniffing airplane glue, sitting in my room with a brown paper bag over my nose.

I was also hanging out with my mother in neighborhood bars. One is still there, at the corner of Magazine and Napoleon, "Ms. Mae's." In the 1960s men who worked on the wharves that ran along the Mississippi River from the Irish Channel to Audubon Park came to the bar for its cheap beer. The owner kept an icebox filled with boiled crab, shrimp, and bright red crayfish, and I suppose I had my dinners there nursing a root beer or a creme soda as Mom got drunk. Many of the men were veterans from Korea or World War II; the younger ones were back from Khe Sanh, Saigon, or someplace else in Vietnam. Except for Mom few women came to the bar except as tattoos—Rita Hayworth, Marilyn Monroe, and other pinups lounging along the men's tanned biceps.

Then there was a wee problem with explosives. With a small arsenal of fireworks, especially cherry bombs, I tried blowing up the neighborhood. Cherry bombs were delicious explosives, more powerful than they are today, small and round with a good wick that left your hands smelling of saltpeter and your heart thumping. Best of all, they were waterproof. After a storm you could throw them into a puddle and imagine great World War II naval battles. Or you could drop them through someone's bathroom window. Kaboom!

Armed with cherry bombs, bottle rockets, Roman candles, and strings of Black Cat firecrackers, I was invincible—until a policeman chased me down. I made it about halfway up the backyard fence. The policeman dragged me to the car, ran his hands down my pants, pulled out a knife. I wasn't cut out for life as a hardened criminal. It didn't take more than a second before I ratted on my friends. Soon there were three of us heading downtown.

Mom was at work, or in a bar. One of my sisters was the first to get home. I sat with my accomplices in a juvenile cell downtown. For some reason the authorities decided to release me with one of my cellmates. I sat in the bed of the pickup truck bumping slowly along Uptown's evening roads, looking through the cab window and watching father and son talk. He had a bottle in his hand, wrapped in a brown paper bag. Father and son passed the bottle between them. I guess he was proud of his son, that being thrown in the clinker marked some rite of passage. At some point, the truck moseyed into a parked car. I jumped out and walked the rest of the way home.

I know now that there were two moments in my childhood when I almost became a ward of the state. In

Louisiana this usually takes some effort. Social services have never been especially robust. My behavior must have exasperated my mother and especially my grandmother. Ending up in a foster home or worse would be a final stain—the unequivocal mark that our family had collapsed, that in some basic way it no longer really even existed. The shame of failure can be especially pronounced among white Southerners, where the past is always a better place, and is just around the corner. We had our stories of the good days, of solid houses and nice clothes, but most of all of familial stability, civility, decorum. Foster care was unacceptable.

I suspect sexuality explains why I headed out to California, as well as the peculiar silences I have encountered during my research. Siblings can recall particular crimes and misdemeanors even as they disclaim knowing why I left New Orleans. A patina of apprehensiveness settles over their recollections, a slight change of voice, a reticent look, as if there are memories without language lurking beneath the veils of consciousness. It seems they are withholding something, less by a willful act of dissimulation than through discomfort, perhaps even an inability, given their own sexual histories, to give voice to shadows.

Sexuality and childhood together are so taboo, especially when involving an adult and a child, that it seems impossible ever to suggest that the two might coexist, that sex is a central feature of childhood and, sometimes, a terrible part of life. My sexual childhood remains verboten and unknowable, though I wonder if the specter of my becoming sexually deviant far outweighed my childhood shenanigans. What is clear is that homosexuality remains for Catholics in my family the greatest ignominy, the quintessential family

failure. And I was effeminate, played with myself, was surrounded by women, and the only man in my life spent his
time parading around his apartment in a tight Speedo with
his semi-hard prick pointing heavenward. Blowing up the
neighborhood was one thing but becoming gay was something altogether different.

———

Flying to academic conferences I watch flight attendants plying a young child's loneliness with drinks and games and
think about what it must have been like saying good-bye to
my mother in the New Orleans airport or seeing my father
at the LAX terminal. Research tells me it was sometime late
in the summer of 1970, when I was ten years old and about
to begin fifth grade. I realize, as well, that the journey was
a sociological statistic. I was one of thousands of children
boarding buses, trains, and planes shuttling between parents, at a point in American history when nearly three quarters of marriages ended in divorce.

I spent nearly a year with my father in Southern California, first in Burbank then at the bottom of a La Crescenta
foothill in an apartment complex near the supermarket and
a tangle of roads. At the end of the hallway there were two
small beds, a closet, chest of drawers, and unadorned walls,
a bare room meant for a visitor. I arrived with just a few
clothes. I left with little more.

My father was in his late fifties, with a shock of thick
hair that had returned to white now that he had a job. Sabrina had gone back to New Orleans. His marriage to
Dorothy the Witch had collapsed. He had long tired of chil-

dren. California had promised escape, a chance to begin again. He wanted a woman, preferably with some money.

The summer I joined him, Dad met Letha, a widow with a house in La Crescenta and a garden of bright flowers and lush foliage and a lawn that glistened due to her dutiful summer watering. Letha was a Southerner like my father, though California had rinsed away most of her Alabama accent. For a lonely decade during her forties and early fifties she had cared for her husband while working as a secretary at the community college in Pasadena, adjusting his pillows and oxygen mask as he lay in front of the television suffocating from emphysema. Letha was a kind, bubbly woman, as eager as my father to begin again. They may have met at a square dance. Their romance unfolded to "take your partner" and "do-si-do" and "give your partner a twirl." Dad bought a cowboy shirt, pants, and boots. Letha wore pink-and-blue-checked skirts adorned with lace over full petticoats. Twice a week in the evenings, and sometimes on the weekend, I sat on a folding chair watching them dance to Western music spilling out of the school hall speaker.

Letha's house sat up the hill from our apartment. On the weekends I slept on a sofa bed, in the mornings watching cartoons while Letha pranced into the kitchen in a negligee and pink panties with lots of frills and a broad smile. During the week Dad and I subsisted on Morton pot pies and TV dinners, and in the mornings on Froot Loops and milk. I spent a lot of time by myself. I learned how to skateboard. My middle finger went numb from playing a Duncan yo-yo, spinning it wildly, trying to get the yo-yo to sleep or to walk along the ground. I stared, dumbfounded, at my father's pornography. He had a magazine about a nudist colony, people standing about naked, men with flaccid penises and women with large, droopy breasts, all very peculiar compared to my experiences of watching women stripping in the French Quarter, with its smells of tobacco and beer and sex's musk.

Most of the week I was bored, stuck in the apartment and in California's cement suburbia. In New Orleans I could bike around Uptown, riding high atop the levee that sat like an ancient Indian mound protecting New Orleans from the Mississippi, pedaling all the way to where the river turned north to Carrollton and the Black Pearl. I could stop for a while and watch the brown water form eddies and carrion trees float downstream, while tugs pushed barges toward the Irish Channel. Or zoom down Monkey Hill, then bike across Audubon Park for a spearmint sno-ball on Plum Street. There were neighborhoods to explore, filled with large homes of bright white clapboard and wrought iron fences and light dancing from cut-crystal doors polished clean by black servants. Reds and oranges turned to blues and purples. Fires leapt and cooled in the corner of my eye. I could take the bus downtown to the French Quarter, floating along with the crowds watching women in their moonlit skin, or I could go to the Everything Store and wander among the curios: New Orleans Saints shirts, velveteen toys, and Voodoo dolls, while children rolled a dime into a game and tried maneuvering silvery claws toward dozens of stuffed elephants, giraffes, and monkeys. In California, I just stewed.

California was a hiatus; I was a temporary sojourner in my father's new life. The school principal told Dad that I was spending far too much time alone. I was neglected, troubled. If the situation persisted the school would have to call Social Services. Something had to be done. It was time to return home.

—⟋⟋⟍—

I realize now everyone had been sworn to secrecy. In my mind's eye Grandmother appears distant, as if in the year apart we had drifted into different orbits. Perhaps it was because of the secret she was keeping, or she felt that welcoming me back to the apartment would make my leaving more painful, or it was simply old age and life's tolls that explain her diffidence. There were no discussions of school or summer plans, the closet room sat unprepared and unwelcoming. Even on the Greyhound I had no clue that there had been discussions, arguments and accusations, agreements reached, new commitments to funnel money from California to wherever, a decision that I would be moving in with my sister Susan in Ocean Springs, Mississippi.

I had just turned eleven, a scrawny kid with a mop of dark hair who had somehow managed his way through fifth grade. Mom had just turned fifty. My imagining of the bus ride is, and isn't, a memory, more re-visitation than recollection, so that what appears before me seems truthful if also embellished. Tenses shift, the past becomes present again, and suddenly I can see the two of us sitting at a small round Formica table in the station waiting for a sonorous voice announcing our bus. Mom pulls a bottle from her purse. She tells me to get a sandwich. I wander around the station under the fluorescent lights, looking at the people standing at the ticket counters and the buses arriving, idling, and departing for various towns around the South, all the lonely faces in the crowd as if I am in a painting by Hopper. A man now sits with my mother.

When we board, she tells me to sit toward the front of the bus. I look out the window, pressing my face against the glass, feeling the air-conditioning coming through the vents. The highway climbs above the city streets, so I could look at the top of the trees and at the diamond pattern of roof shingles as we passed by, some of the asbestos tiles gray with city dirt, most green, a few salmon-colored and cracked by the heat. In places the hard rains had washed the particles of color away. Away from the city, yellow pines nodded and moss hung from cypresses like a child's kite snagged on a limb. Mom is behind me, drinking and making out with a man she had picked up at the station. She is loud and crude, and I wonder now why someone didn't intervene. All along Highway 10 the two of them drink and kiss, the man running his hands across her breasts and up my mother's dress, until I forget about them and listen to the steady beat of tires and Mississippi's pine stands turn into a green blur.

Until I moved in with my sister, I had lived poor. I had changed schools often, my clothes were usually tattered and dirty, and I was constantly sick with tonsillitis and bronchial infections. I was thin, poorly nourished if not malnourished. I had mostly raised myself and had grown up too fast. I had witnessed too much, lived in a near-permanent state of over-stimulation. My mother was a suicidal, sexual drunk. My two sisters, the quintessence of Southern babes, were brunettes and stacked. There was Bobby down the road, and throughout the city plenty of mischief to get into.

Eighteen years separated Susan and me. By the time

our parents divorced, she had married and moved away. She was a foreigner to me, as if she was descended from another family, another epoch entirely, when our parents' marriage had been a good enough one, and Mom had organized parties and hand-stitched dresses, and Dad had provided for his family and carved roast chicken for Sunday suppers. Susan became a woman before my mother began drinking heavily, leaving home for the Mississippi College of Women where she had studied art, then married a man with sugar blue eyes who would become a fighter pilot. She dreamed of wealth and style and beauty, and of seeing the world with her smart and witty husband. They lived in Southern England for two years, traveled to Europe. She took cooking classes at the Cordon Bleu. They held elaborate dinner parties of pheasant and Beef Wellington. She learned from a neighbor how to collect antiques—fine English furniture in mahogany, oak, and walnut. Joe restored an old green Jaguar from the 1940s. They were a regal couple, and in love. Joe would roar past the house in his F-4 Phantom, dipping his wings in affection.

Yet theirs was a volatile marriage. Both were argumentative and headstrong. Susan wanted more than anyone could possibly provide, certainly more than Joe, who would never make it past Captain after he lost his wings in a crash that broke his back and left him in traction for months. Joe had a pilot's exactitude that few could meet. He knew how to pick on someone's weaknesses, turning his quick mind to cruelty. At the dinner table Joe would reduce me to tears, then tease me for crying, turning all "rubbery-faced," he would laugh. I would struggle to steady myself, force a bit of food down, drink my milk, but the tears inevitably flowed, and I would run away in defeat.

I do not know what conversations Susan and Joe had about my coming to live with them. She wanted a son but feared having another daughter. My gender was clear, though in my family's eyes my sexuality was not. I imagine Joe was indifferent to my arrival; he generally disliked children, who, he thought, had nothing intelligent to say. It was for him something of a repeat history. Six years earlier, when my brother Gus was about sixteen, Susan and Joe had taken him in. Dad had called Susan, asked her to take Gus into her home. Our brother had just a few months earlier broken down the bathroom door and saved our mother from herself. Gus was distressed, shaken by watching Mom's decline into mental illness. Mom moved to Mandeville. Gus moved to Tucson, where Joe was at advanced fighter school, learning how to dance away from MiG jets screaming down from above, breaking the sound barrier in the dead of night and nearly getting kicked out of the military.

There is a photograph of Joe and my brother hunting mule deer high in Arizona's scrub-covered mountains. Susan fed and clothed Gus; it was probably the first time in his life our brother had square meals and something like a routine. She had two children of her own, an infant in diapers, the other child toddling about, and a husband jealous for attention and unwilling to do much to keep the household afloat. Gus came to Tucson a broken child and a bewildered adolescent. He was also functionally illiterate. A teacher took him under her wing and helped our brother begin learning how to read and write.

Six months later Gus was back in New Orleans. It had not worked out. It was all too much for my sister. The ordeal must have been terribly wounding—a poor, virtually uneducated kid who had witnessed his mother's suicide attempt, saved by an older sibling and then sent back to New Orleans.

Gus refused to return to the Chestnut Street apartment. With his best friend Terry, he found an apartment a few blocks away, going to Fortier High School in the day and working jobs to pay for rent and food before heading out to California for a few years. Terry and Gus saved each other as best they could. They managed their way through high school, but both ended up in Vietnam, Gus as an infantry-man, Terry as a soldier loading the dead and wounded into Hueys, then washing away their blood after returning to base. Their lives inevitably brought them back to the city they called home, where they continued borrowing money from each other, drinking beer and smoking cigarettes and dope, shooting the shit, and watching over each other when one of them landed in the local VA hospital.

I was younger than my brother, more malleable. I wasn't dyslexic. With the right discipline, with the right care and affection, Susan thought I could be redeemed. But I knew, too, that I was on spec. At any moment I could be returned to sender. Susan made this clear. Infractions brought penal-ties: too many and I would be kicked out. I lived with Susan as long as I stayed in her good graces, on sufferance. Susan reminded me that I would have to pay my way through col-lege. After graduating from high school I would be on my own, if I lasted that long.

Living with my sister Susan was complicated and dif-ficult. There were also basic things to attend to, beginning with a transfer of guardianship so I could attend public school. This involved my sister Marie, who had been granted custody by the court. Forty years later, Marie still feels guilty she did not take me in. Marie has always been the dutiful daughter and sister trying to repair the world. As an adult, she has rescued a drowning man and resuscitated a newborn child who was, literally, delivered into a toilet. She looks after everyone except herself. She tells me that she had four kids of her own to take care of; I know also that she wanted to go back to get her GED so she could go to college and become a nurse, a dream she had held close to her heart since recovering from polio in the 1950s. What we do not tell each other is that she feared taking care of me. Marie had seen too much. Both of us had. I was a liability, a potential threat to her dreams of a solid, middle-class life.

There were basics to learn, like regular bathing, clean clothes, table manners, going to sleep at a decent hour. Susan brought me to the doctor for my vaccinations, and to the dentist, followed by repeat visits. More than a dozen cavities needed filling. I remember the Gulf sun against my swollen, numbed cheeks while I sat on a wood pier casting my rod toward the horizon.

I also learned the costs of being saved, knowing I had been saved conditionally. Forever looking forward, Susan never asked me about my life in New Orleans, or the time I had spent with my father in Mississippi and then in California. She did not, could not, know what had happened. I was too young to tell her—the language did not yet exist—but young enough to want someone to gather me in their arms.

A silence grew up around my sister's disinterest in my history. She provided a different story. New Orleans was the past that I was not supposed to talk about, that I was, in a sense, supposed to forget, except as a kind of haunting punishment I would be returned to if I did not behave. She never asked about, did not want to know about, my white-trash experiences of hanging out in bars with our mother, the bad clothes and the ill-health, all the sex. There was a choice. Live with my sister and renounce my old life, or return to New Orleans.

SIX

—*w*—

TUNISIAN NIGHTS

"There. There it is. See?"

Susan's finger glided across Africa in a maroon volume of the *Encyclopaedia Britannica*. I stared over her shoulder, unsure what exactly she was pointing to amid geography's jagged lines and country colors floating in a sky-blue ocean.

"There. Right there." Her fingernail tapped the page. "That's where we're going."

TUNISIA ran right across the country, beginning in the foothills of the Atlas Mountains and ending in the Sahara Desert—the land was so small it could barely contain its own name. Tunis appeared next to Carthage, then Al-Qayrawan, Sfax, Gabès, and beyond the vast Sahara Desert with its oases and wadis and the Niger River bending into golden sands. My eye wandered northward to Italy and to Europe, but I was drawn back down to Africa, following the continent past the equator nearly, it seemed, to the bottom of the Earth.

I learned that in parts of the Sahara Desert the sands are as soft and fine as talcum powder. During a great storm the sirocco picks up a piece of Africa and settles it right around the world. A thousand miles away, someone sweeps the desert from their home beneath a blood orange morning sun.

I celebrated July 4 on a beach near La Marsa, tossing myself in the sea and playing touch football while trying to avoid the balls of tar that blackened my soles. The beach was a twisting strip of white at the bottom of an olive-colored hillside, formed as if the earth had tumbled into the sea following some epochal struggle. Compared to the Gulf's flat, muddy waters and indistinct pine stands, the Mediterranean had such life to it, endlessly rolling waves and waters swirling around rocky knobs and a sharp, salty taste that left your eyes stinging and red and your hair matted stiff.

Joe had been stationed to the American Embassy in Tunis, the capital. The Nixon administration considered Tunisia a moderate force in the region, particularly compared to neighboring Libya ruled by the radical and eccentric Muammar Gaddafi. Joe's job entailed advising the Tunisian Air Force, which at the time consisted of a mere handful of planes. We rented a house in Carthage near the sea. The market town of Le Kram lay just to the south. On Saturday mornings we returned with fruits and vegetables and freshly plucked chickens. I would buy sandwiches of tuna, olives, onions, and red hot harissa on bread that left my jaws weak from chewing; they were my Tunisian po'boys. In the other direction Sidi Bou Said with its air perfumed with jasmine, secret courtyards behind ornate doors, and along a zigzag of cobblestone paths, red bougainvillea reaching along shimmering whitewashed walls high above a crescent bay. From my bedroom I listened to imams beckoning from the minarets and the rickety small train chattering its way to the çapital,

Tunis. At the top of the hill an Orthodox church, close by a Roman amphitheater. I tramped past Hannibal's port, imagining ships filled with elephants and soldiers, and through the Antonine Baths a short walk away along a sea that tossed ancient mosaic tiles into the sand and seaweed. Friday afternoons a man appeared at the house, thin with a stubbled face and a fisherman's skin and a basket of fresh sea bass, gray sole, rouget with their large gelatinous eyes, sometimes even a lobster or two, always an octopus slunk away at the bottom. There was a stand not more than a hundred yards from the house with pyramids of lemons and oranges, potatoes and green beans, and leeks and lettuces filled with a fine, gray dirt. We would sit on the veranda eating artichokes, rushing through the leaves to get to the bitter heart. On a small Hibachi lamb chops sizzled over a rosemary-scented fire.

And everywhere history. In Carthage I walked upon the wreckage of civilizations, shards and other bits and pieces of the past strewn everywhere. Some were there by resolute chance, a tile stranded on the shore or the curled ochre lip of a vase discarded a millennium ago, others according to a desire to record, preserve, and supplicate posterity, hoping that some god or some man might one day care. I would learn that Phoenician immigrants created the town in the ninth century B.C.E., turning Carthage into one of the Mediterranean's most fabled commercial hubs, ruled by a Council of Elders and a tribunal Aristotle discusses in his *Politics*; I would read about how Carthage fell to the Romans in 146 B.C.E. following a three-year siege, destroying the city before rebuilding it into one of their most important ports, connected to a vast network of roads and aqueducts that reached north into cork forests and vineyards and south into what is now the edge of the Sahara Desert. There were ancient marble quarries to explore, where columns half emerged from the stone, as if the workers one day had picked up their tools and vanished. In the third-century A.D. amphitheater at El Djem, I became a heroic gladiator wielding a sword before an audience of thirty thousand.

I emerged into memory beneath Tunisia's butter-yellow sun, as if ancient history could somehow counsel a boy's troubles, or there was some solace in knowing the world would not slip into oblivion and despair and that the future lay precisely in the bridge we build between present and past, between life and loss. I was twelve, about to begin seventh grade. Tunisia seems as clear as yesterday: the names of teachers, friends, even of people I never met but only heard of; riding bikes along narrow hilly paths; steak au poivre in Bizerte; couscous near Gabès; searching for rifle shells at Kasserine Pass; a month fruitlessly scanning the early evening sky for Comet Kohoutek as it slumbered disappointingly beneath the horizon. I remember camping near the beach at Cap Bon eating merguez, a spicy lamb sausage, with an old man whose language I could not understand. But we were cold and hungry, I had not caught a single fish from a day of casting into the surf, and the two of us sat by the fire where the sausage crackled.

Now I am staring at the furrows of the man's face, which are as deep as the past. Now I am driving with Edward, his sister, and his mother, a small woman, brunette, with fast, slightly wild eyes. We are headed to Sidi Daoud, a tiny fishing village that is mostly empty except for that one month when the men go out in their old wooden boats with nets that dip into the sea nearly a hundred feet down, much as they have since Roman times. A man in a small skiff spying the bluefin tuna that have traveled southward from Sicily signals the moment when it is time to close the nets. The boats draw closer. Bluefin tuna, silvery white and as quick as sprinters, dart from one end to the other. The nets become the *corpo*, the "death chamber." Now you can see the bottom of the nets, and the water becomes confused and roiling, like rapids, except that you are out in the middle of the sea. The men leap into the *corpo* with their hooks and begin lugging the fish into the boats. The water, churning with the muscular fish and men, turns bright red until all the fish are dead and in the boats, and the blood washes into the sea, flowing toward Libya.

—ɯ—

All these memories, so rich in my mind even now, seem drawn with an artist's palette of colors against the blank canvas of earlier childhood. The New Orleans past appears bleached or gray in contrast, as scattered as my father's photographs in a cardboard box. Only by delving into other people's lives has a semblance of order become possible, a continuous narrative, one year following the next, history a prosthesis for lost memory, and only that.

Neuroscience invites me to consider how this all unfolded inside my head, why one past feels as if I could touch it at any time, while another seems forever just below the horizon of memory. I know now I was saved at a biologically auspicious time. During adolescence the brain undergoes significant structural changes similar to early childhood's neural evanescence. The hippocampi, which are so important to the formation of declarative memory, enlarge considerably, as do other areas of the brain. There is neural growth and both the elimination of cortical synapses and extended myelination, as if the brain were renovating its basic wiring one final time.

Scientists have neat terms to describe these changes and especially the relationship between memory and neural development, such as "long-term potentiation" and "coincidence detection," the ability of certain neurons to create or strengthen connections with other neurons, and the process by which neurons create associations between distinctly different inputs. Philosophers all the way back to Aristotle have speculated that the mind somehow is capable of forming a lasting connection between two or more stimuli. Two things happen in space-time, which we detect and assign an association. Contingency becomes causation. These associations can be exquisite or sickening, the smell of coffee in the morning goes with a lover's attention, the sound of a car backfiring with a comrade's death. Memory's enigma is that it leaves us with the sensation of time passing but also, and simultaneously, with the feeling of traveling backward so that some distant past suddenly seems nearby.

We now know something of how this works within individual cells and even at the level of molecules. The hippocampi are important in integrating information spatially and temporally; in a sense they order experience in space and time. Destroy parts of the hippocampi and rats get lost in a simple maze. Scientists speculate that the hippocampi play a kind of record-keeping role in registering the present and organizing it into the past. Memories get the mark of time. Hippocampal neurons seem especially flexible and adaptive. They are also susceptible to destruction. Severe depression quite literally wounds the hippocampi, leading to all sorts of memory problems as well as difficulties knowing where one really is or just how time is passing. Experiences of what happened an hour ago, the embrace of a loved one, the color of flowers one has tended all spring, a child's smile, fade away as if passing through a weir. Some have described depression as being lost upon a nondescript sea beneath a diffuse light where there is no horizon. One is hopelessly adrift while at the same time there is also that persistent awful tugging as if Earth had acquired a new mass and gravitational field, an unbearable heaviness of being that leaves one lifeless yet filled with pain.

Luckily, this area of the brain retains its neural potential; certain antidepressants help stimulate the changes within the hippocampi, which scientists believe may explain why it takes a number of weeks before one begins feeling their positive effect. The creation of associations, the making and strengthening of synapses, is an ongoing process. Working in tandem with the prefrontal cortex, the hippocampi allocate memory to other areas of the brain tied to emotion, cognition, and language, though we know painfully little about how these networks emerge and the degree of their complexity. What seems clearer is that the process is contingent within a general framework and order, a peculiar and ever-changing combination of chance and necessity.

Human brains display remarkable similarity. Every neuron has pretty much the same features: a cell body, dendrites that receive signals, axons that send them on their way to other cells. Look more closely, say, using an electron microscope, and everything suddenly becomes different. No two dendrites are the same. Little spines protruding from the dendrite often make synaptic connections with nearby axons. Many dendrites themselves are constantly changing. And with this radical particularity, utterly distinctive connections are made, remade, or destroyed—vast differences created within generalizable patterns.

This happens on an extraordinary scale in childhood and in adolescence. There are as many neurons as there are stars in our galaxy. Each neuron might have anywhere from one thousand to ten thousand synapses. In other words, inside our heads are upward of a hundred trillion different connections, a universe of possibility that poses among the most profound challenges to neuroscientists.

Our brains are thus astoundingly complex. But the important point is that they are also historically constituted. The lives we lead are part of us. At birth, our nervous system is very undeveloped. The brain is especially small compared to its adult size. We grow immersed in relationships and surrounded by all sorts of environmental stimuli that influence changes at the cellular level. Recent research on the adolescent brain suggests that in subtle but powerful ways cultural norms may shape neural processes. What is deemed important and the ways in which we are raised as children and adolescents have neurological consequences. History quite literally changes the shape of our brains.

The past makes us, and we make history, though never according to conditions we choose. My mother's suicide attempt is an indelible part of me. But memory endures because of the creation and strengthening of synaptic terminals, so that the past becomes, for a minute or for a lifetime, permanent within our being. Memory is less a photograph

of something gone than a story created and endlessly re-
newed, revised, or forgotten. It isn't arbitrary either, nor
some imaginary concoction . . . and certainly not a facsimile
of what really happened. Like language itself, memory (the
exception being so-called false memories) contains within it
the trace of the real, a fragment of the past that comes to us
like some glass found along the shore, freshly broken and
still sharp, or softened by the sea. We stand there holding
it, wondering from whence it came, what meaning it may
yet hold.

One result is a sense of history, or historicity, and its
tribulations. The past is gone yet we inhabit it. It seems distant, and then suddenly we have the sensation of memories
cascading over us. We begin to feel that we have existed in
a world we did not quite make, and that world resides
within us. As adolescents we compose stories of our selves,
histories of who we are, how we are different from our parents, how we might be tomorrow or many years hence. It is
a way of making experience sensible, discerning our life's
plot, which requires us to think forward and backward. We
observe the world and observe our very selves. We begin
stringing experiences along time's thread, begin telling a
story, claim a world as our own. No wonder adolescence is
a time of memoir, which explains the genre's popularity in
junior and high school classes as well as all sorts of record-
keeping, from diaries and incessant gossiping to school yearbook prognostications.

"This is who I am because of these desires, experiences, memories," we seem to be saying. Amidst all the commotions and disquiet of being a teenager, wanting to be
distinctive but anxious about being identified as different,

an "I" somehow emerges and begins making appointments with the future.

I imagine neural connections proliferating alongside earlier scarring or connections that could have been made but were not. The clarity and richness of my Tunisian memories is so different from the scarce recollections of living with either of my parents. The issue is not that time's passing has rendered cloudy what once had been clear—at least not only this. Twelve years of life in New Orleans, Mississippi, and out West remain irreducibly jumbled. Siblings seem like figures from a dream, real but also incorporeal, the traces of some dark history. I cannot remember much except for those intruding memories I wish I could forget. Facts discovered in adulthood about my young years remain ephemeral. For a short time they sear, and then disappear, leaving me doubly searching for secret stories of childhood lost. I cannot even remember the notes I have taken of my life, names, places, happenings around which I have managed to append a date, sometimes an hour, even a minute, forty years past—histories written and then just as quickly fading away.

———ɷ———

I remember that feeling of becoming me, as if those Tunisian memories comprise the personal property of some willful person, the evidence of oneself. I remember staring at Roman mosaics of gods, lions, deer, and sea creatures during a class visit to the Bardo Museum. I remember knocking on the door of the Soviet Embassy in downtown Tunis. I want to learn about communism, I said. They looked at me quizzically, then invited me downstairs to the library. I left with texts by Lenin printed on tissue-thin paper and glossy magazines of glorious life behind the Iron Curtain. I remember losing myself in the poetry of Robert Frost, Wallace Stevens, William Carlos Williams, in awe that beauty could emerge from the mere ordering of the simplest of words. I remember Rachmaninoff's piano concertos, and Beethoven's Pastoral Symphony, which I can still play in my head and makes me weep every time I hear it on the radio. I remember floating on Mediterranean waters, watching shimmering seaweed beat back and forth, hoping I might see a big fish or at least an octopus slinking across the rocks.

And there were dance parties in the candlelit basement of Edward's house. Most important, there were girls. Especially Diane, French-speaking, sophisticated to my adolescent mind, and, when I could summon up the courage, sometimes willing to French-kiss as we slow danced. I remember holding her in that adolescent tremulousness to the sounds of Santana's "Samba Pa Ti" and the smell of patchouli on her gossamer blouse.

Well-nourished, I grew unexpectedly. My test scores improved dramatically, though I was not an especially good student. In her skirt and blouses with wide 1970s collars, and dark hair slightly parted to one side, Mrs. Guediche helped me with my reading and writing. I took math from a Scotsman who seemed as wide as he was tall. Mr. Hepburn was not to be toyed with. Mr. Clegg, a remarkably effervescent science teacher, let us burn manganese tape under water, make our own hydrogen, nurtured our boyish predilections for blowing things up. I took piano lessons from Madame Momy, who lived just a few blocks away from our Carthage house, an exacting woman forever correcting my posture. I spent an entire year learning scales and endlessly practicing Czerny exercises, until I insisted that I be allowed to play "Eleanor Rigby." She was not amused. Everyone at the school hated Madame Momy and her European strictness, but as our music teacher she somehow persuaded our cracking voices to rise to Handel's "Messiah."

That first year Susan and Joe briefly entertained the idea of formally adopting me. I remember the sting of learning that my father had willingly agreed, as if it was a decision that demanded no reflection. I did not want to be adopted. I harbored, and sometimes still do despite myself, the idea that my mother would heal herself, that in her healing I would someday return, that we would be together the way we were supposed to be, mother and son. Adoption seemed like an irrevocable end. Mom's refusal came as a relief, a kind of rescue, as if from some great distance she was claiming me as hers; but I knew at the time that she was incapable of taking care of me, of being a mother.

I had nothing physical to remind me of New Orleans, which receded to the back of my mind like dark matter in the universe, invisible but powerfully there. Neither Susan nor Joe asked about my previous life. I do not know if it was from benevolence or indifference. It was as if my childhood no longer existed, except in panic-filled nights. From time to time I wet the bed.

"Refuse. Tell her to go to hell," my school friend Victor would tell me after a day of hanging out and watching Susan order me about. "Do something! She's such a bitch!"

Victor reminded me that I was being treated like a servant. I ran errands, cleaned the yard, took the dog for walks, learned how to work in the kitchen. I folded napkins, polished silver, and arranged table settings, memorizing the precise order of utensils, their proper distance from the plate. On weekends I helped with the cooking. I learned how to serve food at elaborate dinner parties. "Serve from the left, remove from the right. And never, ever, stack dishes in front of guests. Plate the women first. Twist the bottle at the very end of pouring the wine."

But I wanted to please. I was a visitor in my sister's family, a proverbial Cinderella who has received a final warning before being kicked out and who wants to be loved so desperately she imagines fairy godmothers and a prince who will rescue her in the end. Boys have no such dreams. They are meant to be on their own. In the afternoons I walked along the avenue punching eucalyptus trees, knowing I did not belong and one day would have to escape. Susan reminded me that while I might stay with them through high school, after that I would be on my own. I should expect no support. But these seemed reasonable costs for being saved. They had, after all, done more than their fair share.

I wanted to interview Susan and Joe all these years later. Both refused numerous requests, the only members of my family to have done so. They are long since divorced. Susan answered a few basic written questions of fact, but insisted that she not be quoted without prior permission. It is easy to pass judgment. Is love possible when it is dispensed on sufferance, as a quid pro quo? "You bastard," she would yell at me, "I'll send you right back, this instant. I saved you. You had nothing. Nothing. Don't you forget." She was also a woman barely in her thirties, in a difficult marriage, with two children of her own, who had taken me into her home.

Joe returned to Mississippi. I visited him recently. We are on cordial terms. But when I began reminiscing about the past, gestured toward a question, he turned sarcastic and then changed subjects. Joe was not interested in being a father, not to his own daughters much less to me. On the weekends he would have me watch him work on his Mercedes. He commanded me not to talk. I would hand him tools, a socket, a screwdriver, panicking as I rifled through the box to find the correct wrench before he began berating me. In the second year Joe bought an old BMW, a single-cylinder motorcycle that he struggled to start. I thrilled holding onto him as we rode around Carthage.

I wanted to ask Joe about fathers and fathering, about how I might have seen him in those years I lived under him. He was not much cut out to be a father, particularly of younger children. He had a fighter pilot's meticulousness, which would not tolerate child's play and disorder. Exposing his faults or castigating him was not my intention. I simply wanted to know Joe's observations, what evidence he might be able to summon and share. Was it he who gave me a copy of *The Old Man and the Boy*, or was it a teacher who kindly observed my need for a father, even one found within the pages of a novel? I have since returned to the book, purchasing an exact copy of the text I read in that Carthage living room, with its first words "The Old Man knows pretty near close to everything." Also, "The thing I like best about the Old Man is that he's willing to talk about what he knows, and he never talks down to a kid, which is me, who wants to know things."

I wanted a father, or, like the fatherless boy, a wise Old Man who would take me hunting, instructing me in the ways of quail, pheasant, and mallard, of staring down that first deer in an autumnal wood, of bringing a shotgun below my chin right and proper. I would learn about love and respect, and when I walked one day through the woods alone I would do so as a man. Re-reading *The Old Man and the Boy* and the *Old Man's Boy Grows Older*, I realize how much I learned from this simple text of hunting and friendship, instilling in me a belief that I could learn anything from a book, anything and everything. I remember crying uncontrollably at the end of *The Old Man and the Boy*. The old man and the boy had returned from Johns Hopkins, where I would study a decade later. "I ain't got to tell *you* that I am going to die . . . You've had the best of me, and you're on your own from now." I kept the book for years, refusing this end.

"It's very good," Mr. Liska tells me.

I am in the headmaster's office. He is a kind man, I can tell, with black glasses and a square, John F. Kennedy face, though the vast expanse of his desk and the fact that I am sitting in the headmaster's office, door closed, has me trembling. But Mr. Liska is patient and speaks calmly, trying to assure me that I am not in trouble, and that we are here simply to talk about why my story has been rejected for publication in *The Carthaginian*.

"And I am glad you wrote it. But I don't think we can publish your story. It might be a little too scary for some of the younger children to have it in the school yearbook."

I am not sure I understand. Mr. Liska's voice seems disembodied, the words floating about the room, which has begun to feel like a prison cell. I am convinced I've done something wrong, feel trapped and ashamed.

I learn Dr. Liska has spoken to Susan describing the story I wrote in a sloppy hand on the pages of a spiral notebook. There is a boy, a father, a gun, a hunting trip, a child's electric excitement lying in bed trying to sleep. They would wake well before dawn, drive to an open field on a hilly expanse of farmland to hunt quail, just like one of the stories in *The Old Man and the Boy*. But somehow while dreaming the boy gets out of bed and rigs a trap. When the father opens the door, the shotgun explodes in his face.

Joe did take me hunting one day, for partridge on an old Tunisian wheat farm. I remember struggling to fall asleep the night before, and driving through the dark in the old green Mercedes. There was a heavy dew on the ground that had begun turning to a mist, rising a few feet. I walked behind Joe, with his sawed-off 12-gauge pump-action shotgun, hoping to see birds foraging for loose kernels.

I cannot fix, either in my mind's eye or by research, the sequence of things, except that I wrote the story after learning of my father's willingness to have me adopted. Or perhaps it was in some sort of protest against Joe, who could be cruel, laughing at my tears, my rubbery face. The story I had written seems as real as the dirt clinging to my shoes that day walking on the farm. Nor can I discern the swirling disappointments within me, except that I know, now, that I was mourning and angry about lost fathers and looking for an Old Man.

There were not many men I had grown up admiring. There was, however, Pablo Foster, who once came to school. Pablo sat on the school board; along with others, he helped create a curriculum that folded Tunisia's miasmic history into the intellectual habits of restless children. He was to give us a lecture on the history of Tunisia. We gathered in the small library, fidgeting in our chairs, some of us making spitballs. There was a map, held up on an aluminum tripod. Fluorescent bulbs gave the room the sickly white light of a doctor's office. No one wanted to be there. We wanted to be out on the field playing soccer.

A teacher introduced Mr. Foster, with his beaten-up brown shoes and runaway black beard. He stood up from the folding chair. Not more than a minute or two into the lecture he somehow managed to bump into the tripod, which came magically crashing down. Pablo picked up the tripod, stumbled about, tried repositioning the map, but whatever authority this adult had had vanished.

"Annnnnndnoooooow. For my next trick . . ." I turned around and looked at him. He wasn't embarrassed at all. He had somehow managed to recapture the attention of a bunch of thirteen-year-olds, or at least mine. Pablo proceeded to tell us about Tunisia's history, not just the Phoenicians, Carthaginians, and of course the Romans, but also about the culture of the people of the desert and of the Sahel, the great trade routes across the Sahara Desert, the camel caravans traveling from one oasis to the next. He was able, just with words, to conjure in my mind images of a world that was utterly foreign yet somehow right here, in a simple room.

There was something magical about this fumbling man in a school library, fluent in Arabic, who seemed to know everything. With other teachers I was accustomed to their constant disciplining. Pablo was different—a teacher, but funny, a little aloof, wise, and slightly out of control, with a crazy beard and uncontrolled hair in a world of clean-shaven men in suits and ties.

Later that year my sister, nieces, and I traveled with Pablo and two other families straight down to the edge of the desert. Pablo showed us the mosque at Al Kairouan, one of the most outstanding examples of Islamic architecture and a revered site for Muslims. We went to Gafsa and as far south as Gabès, then back north to Sfax. A short ferry ride brought us to the Djerba, the Island of the Lotus-Eaters. Here, according to Homer, the sirens' melodic sounds had so seduced Ulysses' sailors that they jumped off their boats and drowned. Pablo brought us to an ancient Jewish synagogue, the oldest in all of North Africa, its inside cool and dark. There were Roman ruins to explore, World War II battlefields to cross. In an oasis I climbed to the top of a palm and with a thousand drunken bees watched the sap flow from its exposed heart. We tramped through alien Matmata with its cave rooms that I would see many years later in *Star Wars*. We ate camel, spoke to Peace Corps volunteers, beat the Sahara from the car following a sandstorm, slept in the desert beneath a billion stars.

It was a short trip, not more than a week, but it changed my life, although not in any immediate way. I fell in love with Africa, and with the past. I wanted to be like Pablo, educated and filled with facts. Pablo visited a few times. I waited on him and his wife, wondered at his exuberance and intelligence and sparkling dark eyes hidden beneath all that hair. Although Pablo drifted from my life, I remember that trip and his fumbling lecture, and I discovered not simply that the world was a very big place, and that one can be saved by education, but that possibility lay somewhere in the distance, just out of reach.

SEVEN

―⚮―

FAR AWAY

A MAP TELLS ME THAT WE MUST HAVE DRIVEN DOWN Georgia Avenue and along the beltway to the Washington National Airport, past leafy Maryland suburbs and the nation's capital filling with summer tourists. I don't recall anything about the trip, the conversations that left me at the terminal that June 1976, the timbre of our good-byes, the way her face met mine, the instructions adults deliver to kids boarding planes alone. I suppose sometime that spring Susan had said I needed to visit Mom. It had been nearly five years. Time to see her again. I don't think I complained. It would be a short visit. I was sixteen and about to begin my junior year in high school. I'd soon be back with my high school friends, throwing Frisbees in a park, mowing lawns for pocket money, counting the days until we could zip around the neighborhood with shiny new laminated driver's licenses in our pockets.

Two years earlier we had packed up the Carthage house and boarded a TWA jet to the United States, moving first to

California and then to the East Coast. Joe had a desk job at Victorville Air Force base in the Mojave Desert, filing papers into sheet-metal cabinets that rattled each time a fighter jet tore down the runway. We rented an A-frame nestled in the bone-dry San Gabriel Mountains, before moving into the valley to one of the small houses reserved for mid-level officers on the base. In the summer I played tennis and fished for silvery rainbow trout. A paper route paid for lift tickets at a small ski resort. I had an old pair of lace-up boots from a school sale, and six-foot-plus wooden skis that towered over my slight frame. A saw reduced them to a more manageable size.

California and Maryland had some of the best school systems in the country. A bus took a classmate and me to the nearby high school for an advanced mathematics class. I spent weekends scouring the mountains to catch collared lizards for my biology teacher, a short, exuberant man and amateur herpetologist. I helped him care for the collection of reptiles that ran from one wall of the classroom to the next: lounging green iridescent iguanas, tegus, a small monitor lizard, and snakes, including a few copperheads sunning themselves under a white light. He taught us how to conduct field ecology. Twenty or thirty teenagers moped about the flat desert creating censuses of the fauna and flora, all of us hoping we might find a rattlesnake hiding near a rock, anything to bring some drama to our scientific work. Some of the mice he raised at the back of the classroom developed tumors. I proposed an independent research project that entailed euthanizing the mice, removing the tumor and staining the tissues on a half-dozen slides, then standing in front of the class describing my findings and lecturing on the genetic basis of cancer.

My history teacher in Maryland took me under his wing, suggesting I read William Hardy McNeill's *The Rise of the West*, an eight-hundred-page tome I lugged around for months. I learned about the world's religions and the creation of what McNeill called "ecumenes," intercommunicating zones that knitted together the most intimate aspects of people's lives—a religion, a simple crop, a treasured spice from some distant land. I found myself drawn back to a map of Africa and my Carthage days, and from there wandered into complicated arguments about the rise and fall of civilizations in the interplay of "steppe" and "sown" peoples, first raised by the great scholar Ibn Khaldun in the fourteenth century.

In December 1975, we left Joe in a dash across America. In the selfishness of adolescence and comfortably lost in the American middle class, I hadn't realized that Susan's marriage had been crumbling for quite some time. I had taken for granted the screaming fights and slamming doors, a voluble relationship scarcely quieted by Sunday afternoon naps. I didn't realize Susan was taking the first step in a new life. I thought it was just another move that would bring me closer to friends I had made in Tunisia.

It's difficult discovering what exactly happened, harder still glimpsing the inside of any relationship. Except when they enter the legal system of lawyers, judges, and social workers, domestic dramas seldom produce much of a documentary record. This type of data rarely discloses what unfolded, which helps explain why the history of the family presents such a daunting task to scholars, let alone to those who go looking into their own pasts.

Oral history presents its own difficulties. Stories of

personal experience can exhibit a powerful obduracy, the past claimed as private property. The self may be a product of a certain narrative historical imagination, but we are loathe to recognize it as something of a fiction. Instead we resolutely hold to certain pasts as our own because *that* past is who we *are*. We stick to our guns. This tendency may be especially pronounced in families where there has been significant turmoil, making it difficult for the emergence of new narratives that might offer some succor to nagging wounds, a historical awareness that heals. Troubled families are all about silence and secrets, the quiet despair behind Christmas Day smiles, the things that remain unsaid even in the most intimate accounting of shared pasts, all the forgetting upon which memory is made.

Possibilities for a different recounting of past nonetheless reside among the artifacts families keep, the rituals enacted, the reasons family members insist *this* happened for *that* reason, even the silences and aporias that reside within personal narratives. In my case, it's been difficult trying to gather some of the pieces. My father passed away nearly twenty years ago; I am left sifting through a small box of photographs and assorted records. Mom offers the terse "I can't remember" to many questions, no matter how I try draping them in soothing language. In the middle of a question she simply turns away. Siblings offer a kind of vacant response to my queries, the sort of "I'm not sure, what do you think?" answers that leave me wondering. Susan and her ex-husband rejected repeated requests for interviews. They seem afraid, as if I am gathering facts for a prosecution in a case half a life ago. Or perhaps that I will expose dark misdeeds committed in a marriage's unraveling.

Amid the silences and silencing research has produced a few facts. Information has a way of leaking out. The world is just too messy, too complicated, for any one of us to contain or sanitize the past, to hermetically seal history the way we might wish, no matter the costs. Subjugated pasts surface in the most unexpected ways: a document an official forgets to shred, a slip of the tongue, even what remains unspoken.

I discover, for instance, that I boarded that flight to New Orleans in the early summer of 1976 with a one-way ticket. I wasn't meant to return. Susan was sending me back to New Orleans, hopefully for good.

I email my sister, trying to abide by a standard rule of historical research: validate what you've found. She refuses to say anything about it, or to have her words reproduced here. My question about a bit of information that invariably concerns an intention—the "why" behind every human act—opens a thunderclap of words about how she saved me, how terribly neglected I had been, how under her care I had received medical attention, decent food, and an education. I want to say that all of this is true. It is true. I am forever changed by her kindness, and thankful. I owe her my life. Had she not saved me I wouldn't be where I am today. Most likely I'd be poor, troubled, and miserable. She did more, much more, than anyone else—more than I would have if I were in her shoes. But it's impossible to have this conversation. She interprets my question as an accusation, pleads a familial Fifth.

A simple fact, a little piece of history we happen upon, changes how we explain the past. This is often how it works in scholarship. The historian knows the end of the story. A war breaks out. Politicians sign a treaty. Peasants revolt, laborers strike. A kid boards a plane to see his mother. Something happens. Historians are reverse engineers of the past, gathering material and trying to figure out how the pieces fall together, realizing how much remains unsaid, unknown, or lost. It's the weighing of evidence that's so difficult, the meaning we attach to what we find while rummaging through the lives of others. It's here that the historian becomes a storyteller. This is why a newly found or reconsidered fact can be deliciously subversive. We have to tell the story all over again. The past begins looking different, as does the present.

We do this in our own lives, often subconsciously. We are constantly altering the story of our selves, using information (however flawed or incomplete) to explain what happened a minute or a decade ago, making various assumptions about what we and others were thinking at any given point in time. Interminable revisionists, we create and recreate our lives through an endless process of addition and diminution. Usually the changes resulting from new information are very small, a shift of emphasis, a slightly different hue to our remembered past as we fit new information into the script of who we think we are. Or they can be tectonic, altering in some fundamental way how we understand past relationships and who we have become. Why did person "x" do thing "y"? Why did we act the way we did? Would we even like the person we once were?

The difference with the professional historian is that we rely not only on the found material fragments of history but on our memory, with all its frailties and insistence. Wonderfully retentive and utterly bewildering, resolute and feeble, memory is always already an interpretation that usually contains within it something real, a trace of the past that somehow still feels present. We cannot survive without the ability to summon memory with a certain general degree of accuracy, either implicitly or explicitly, for it is by memory that experience becomes knowledge, whether this relates to the naturalness of riding a bike, discussing what we did at work, or recalling that a certain fishing lure works particularly well in late summer afternoons. What emerges in our minds is not an exact replica of what happened but rather a complicated (and to our minds utterly persuasive) story assembled around a "fact." Memory is a kind of fiction that's usually not totally true.

Usually, because sometimes we think we did something that we only imagined doing, or we fill in what we don't know around what we think we do. Memory, as innumerable studies have shown, can alter by mere suggestion, a kind of editing of the past that can unfold unconsciously. And of course memory can weaken over time. I will be able to recall what I did an hour ago. Six months from now it will be lost.

Because memory is processual, it is subject to revision in the ongoing telling of our selves through which we build upon the likenesses of that which is already inescapably lost. This internal process of revision unfolds surreptitiously, however, in part because memories very often appear in our consciousness unexpectedly and through a web of association with other emotions and sensations. Tastes and smells can awaken the past, most likely because these senses are connected to the hippocampi, which play such an important role in autobiographical memory. This may help explain, for example, why I don't go to bars much. The smells leave me in a funk. Somehow the particular combination of tobacco and hard liquor, especially cheap American whisky, creates a slight queasiness. I can feel a once distant yet still proximate past returning. The present becomes a little heavier. My vision turns a pixilated gray. I'll become sad without knowing exactly why, aware of a cascade of bodily sensations without explicit remembrance. Then unfortunate moments tumble into awareness. I end up back in New Orleans with my mother trying to kill herself.

—⁓—

In the summer of 1976 when my one-way flight arrived and my sixteen-year-old self slouched out of the plane, I could see her standing there uncertainly at the end of a tubular ramp that bent like an elbow from the plane to the terminal. I kissed her cheek. I could smell the booze and mentholated Salems. There were no questions or arms thrown around me, no smile. Mom was drunk, though at least standing. I carried my Samsonite into the wet heat of the airport's exit, stepped into a taxi, headed to Chestnut Street. That evening I sat on the cot in my old room, and looked to the door my brother had knocked down all those years ago.

Mom stole away to the primordial loneliness of her bedroom, with its piles of clothes on every flat surface and the dank smells of stale beer and cigarettes abandoned on a saucer, every one of them crushed and kissed red. Uneaten food lay strewn about the kitchen, a pot of Campbell's Chicken Noodle Soup on the stove, a chaos of dishes in the sink, and scattered across the room the opalescent egg casings of cockroaches.

Grandmother had been dead for five years. Sabrina came home one day to find her crumpled in her wheelchair, her head resting on the desk. The ambulance whisked her to Charity where she died a few days later. Mom left the front room empty, not a shrine to her mother but rather an indifference born of their troubled history. Everything was in order. From the small table that had served as an altar, the Virgin Mary watched over the tightly made bed. It was as if Grandmother was still there, about to wheel herself in from the kitchen, the gold and garnet rosary settled on her lap.

Mom now had neither Grandmother's Social Security check nor the child support my father had irregularly sent her way. She barely survived, never secure but not quite destitute, always a few months behind on her bills. "They got enough money, too goddamn much money," she liked to rail. "They don't need mine." She felt entitled to their services free of charge. Bills seemed a personal insult, even a conspiracy. Mom would wait until the utility company or Bell South threatened to turn things off before she stuffed a Hibernia Bank check into the envelope.

Mom had just turned fifty-five, with a mouth of dentures and bridges and blackened teeth, and a face battered by smoking, booze, and despair. She conjured masks of heavy foundation, rouge, thick lines of mascara and red lipstick, which she removed with Pond's and toilet paper—except on the weekends, when she fell asleep drunk and woke the next day with blacks and reds smeared across her face. The Magazine Street bus took her to a five-and-dime half a block from Napoleon Avenue. She stood at the back selling Lee jeans, cotton-polyester shirts, and Fruit of the

Loom underwear to men and women who lived in poor working-class neighborhoods near the railroad tracks that ran alongside the Mississippi. Lunch brought her to the corner bar where she smoked and drank a few beers and snacked on Saltine crackers. She'd return to the bar after work for a few more drinks before stepping onto the bus back home, sitting next to men roughened by war and by loading containers from ship to train, strong men whose flesh had softened from booze and youth's end, with large tattoos drawn down their arms of big-breasted women in bathing suits.

My brother Gus lived on the corner not more than fifty yards from the apartment, renting a small gloomy corner house with a chain link fence. He slept on a mattress thrown on the floor, surrounded by dirty clothes, empty beer bottles, rolling papers and the last seeds and twigs from a bag of dope. Garbage cans that had missed the weekly pick-up overflowed with rotting food. Roaches dashed everywhere. Tools spilled across the cement driveway.

Gus had returned from the Vietnam War a few years earlier, spending the days hiding in darkness behind masks and goggles and heavy clothes, popping on acetylene torches, powering up diesel generators for arc welders that snapped angrily and covered him in sparks and a sharp white light. Terry had returned from Vietnam too, along with many of Gus's boyhood friends who had been drafted into the war. They had long since burned through whatever money they had on cars and motorcycles and drugs, growing their hair out and spending afternoons and evenings sprawled across couches and beds.

Gus boarded a plane for Vietnam just as I was receiving the first decent education of my life. He landed in Southeast Asia around the time of the Easter Offensive of 1972, when twelve divisions of the North Vietnamese Army invaded the south on three different fronts. The strategically important city of Quang Tri fell to the Communists. The entire coastal area became a "highway of terror"; a fantastic amount of U.S. firepower fell on the country. B-52 Stratofortresses lumbered their way into the sky from Guam; there were so many of the great planes they didn't know where to park them. Hundreds of fighter-bombers departed from across the region, a total of nearly 60,000 sorties. Over 100,000 people died.

Gus had a nervous breakdown. It's a part of his life no one will discuss, especially my brother. I know not to push. All I can discern is that he had been drafted into the war, something dreadful had happened, and Gus had ended up spending a good bit of his tour confined to base before disembarking at the Oakland Army Terminal and making the short trip to the Letterman hospital in San Francisco. It was serious enough that Mom flew out to be with him. The doctors patched him up and sent him on his way.

I had seen him for a brief afternoon in the summer of 1974. We had just returned to the United States from Tunisia. Gus arrived on the back of a motorcycle. He had been discharged barely a year, the physical traces of his military service lost to a wild beard and a black sheen of straight hair. In my pubescent selfish innocence I had no idea what he had endured, and worse, no interest in knowing.

We stood outside the door.

"Dontcha know what a bear hug is?"

I didn't.

"C'mon brother."

Gus grabbed me and didn't let go. We stood there, the smell of Lucky Strikes in my brother's hair. He held me for what seems an impossibly long time, finally let go, smiled, and said something like I was turning out okay, kicked his bike on, and left.

When I returned to New Orleans two years later, he had his hair cut back, though his face remained hidden behind an unruly beard. We fished a couple of times for the sheepshead that lurked near pylons sunk into Lake Pontchartrain. I learned how to smoke dope and drink beer. The weeks turned into a kind of muddy haze. During the day I worked for Louis, my Aunt Cecile's son, scrubbing stone with a solution of muriatic acid in newly built suburban homes with large stone fireplaces that seemed out of place in semi-tropical Louisiana.

In the late afternoons I walked up and down Uptown streets, often toward Audubon Park though sometimes to Plum Street for a sno-ball. Mockingbirds chased one another among red and white pinwheel flowers. Air-conditioners half hanging out of bedroom windows snapped on and off. Back in the apartment I disappeared into my closet room, falling asleep to the oscillating fan click-clicking back and forth and the air sweeping across the sheets.

For the Fourth of July weekend we piled into my brother's pickup and headed out to a small cabin in the woods somewhere up in the country across the lake. In the still afternoon and into early evening, trucks and old station wagons lumbered up the dirt road and abandoned themselves around the cabin. The sounds of Triumphs and

Harleys thumped through the pine trees. On the screened-in porch, men newly returned from Vietnam drank and smoked, laughed and argued. I met a woman named Valerie, who arrived on the back of one of the motorcycles in shorts and a sleeveless shirt, her hair cut short, large breasts loose behind a thin cotton shirt. The men seemed foreign to me. I gravitated toward Valerie, who was ten years my senior. We talked through the afternoon and into the evening, the beer turning youthful confidence into boastful exaggeration. I described living in Tunisia, quoted from books and spoke French, whatever I could do to demonstrate my sophistication and, most insistently, the absolute importance that we have sex.

My seventeen-year-old self could not summon the wisdom to say that this was all too much, that the last place on earth I should be was in New Orleans with my drunken mother, stoned on pot, drinking alcohol, having sex with a very experienced woman. I didn't call Susan to ask her to take me home. I didn't call anyone, even if somehow I knew that I had to escape. When I heard that two of Sabrina's friends were heading to Key West I joined them. Anything was better than New Orleans. Anything was better than being with Mom.

Sabrina worked as a waitress in one of the town's many seafood joints. I had no money and barely any clothes. We slept in a shack at the back of a house where there was an avocado tree. I would climb the tree and drop the green fruit into my sister's hands. In the afternoon I biked around the city selling avocados for a quarter. I made enough money for a lunch of yellow rice, black beans, and a piece of fried fish. In the evenings I ate dinner with the Hare Krishnas at the end of Duvall Street and watched the silver tarpon flash through the clear water.

—⁂—

Returning to New Orleans had been a disaster, enough of a calamity that I had left within a matter of weeks. Susan technically had guardianship, a legal responsibility. Perhaps the idea that I would end up as a dropout in Key West was too much. There must have been some sort of conversation between Mom and Susan, followed by phone calls to Key West. Susan must have changed her mind. Sabrina and I hitched the thousand miles north.

Within a few bewildering months I had taken that plane to New Orleans, hitchhiked from Key West to Maryland, smoked dope, lost my virginity to a much older woman, tossed back shots of tequila until I passed out. Back in Maryland I dropped acid, ate hash, smoked PCP, and worked nights and weekends in a shopping mall making pizzas to pay for dope and booze for weekend parties.

Lots of middle-class white kids did the same, particularly in affluent Montgomery County in the permissive 1970s when the state had all but decriminalized possession of small amounts of marijuana. Drugs suffused my high school, with students splitting bags of dope in the stairwell and getting high during lunchtime. I may not have yet become the stereotypical troubled teenage boy, but I was certainly skating on thin ice. My relationship with Susan deteriorated. Just after my seventeenth birthday I ran away. Or I was kicked out. It's a little hard to tell, even now. I was *going* to get kicked out one way or another. I was *going* to end up with my father back in California, or in New Orleans with my mother, or homeless, or in a reform school. So I ran away in some sort of adolescently unrealistic preemptive escape.

Joe had just returned from California, newly retired from the military and hoping to repair his failing marriage. I had been keeping a diary, I told him. At dinner one night Joe began asking questions in the biting way that had once turned me into a puddle of tears. He had read the diary and its details of teenage misdeeds. I cannot reconstruct the exact words, only the feeling that evening sitting at the table, knowing that somehow the world was dissolving around me. I got up, dashed downstairs to my room, and tore out of the house, running through Glen Echo and Bethesda, Maryland, past three- and four-bedroom homes of brick and ivy set comfortably away from the streetlights shielding the neighborhood. Doctors and lawyers and psychiatrists lived here, and tens of thousands of government employees. I had lived here too, however recently, however provisionally. Their children attended Walt Whitman High School or Bethesda Chevy Chase in one of the nation's very best public school districts, or went to tony places like Sidwell Friends. Good schools meant admission into solid, established colleges and universities, ultimately a professional degree, and an expected return of those sons and daughters to successful careers and protected lives in one of the subdivisions curling out of the Boston–Washington corridor.

Susan called Dad and began packing my things into the Samsonite. The high school principal arranged a liaison. Dad would pick me up and whisk me away. The question was, where? Nearing retirement, happily married to his third wife, the last thing he wanted was a teenager in his house. I didn't want to return to Los Angeles, I think I told him, where I had been miserably lonely staring at the television or ambling through the apartment complex. California was a continent away from my friends. I wanted to stay where I was, finish high school, go to college, then maybe to law school. I would somehow end up successful, secure like all the other white middle-class kids I knew, far away from my New Orleans childhood.

My father and I flew south for the simple reason of figuring out where I should go. His goal was to dispose of me in some reasonably suitable way. I was too young to live on my own, so abandonment was not really possible. Ending up in foster care would mark us as white trash, violating some unwritten mythic rule of the lower middle class and its pretensions to familial fidelity. Something had to be arranged.

Mom never asked me to stay with her. I suspect she refused when Dad popped the question. Or my father, looking at the stack of dirty dishes and the roaches and general dishevelment, sought an alternative, realizing he'd be back on the plane within a few months trying to pick up the pieces the best way he could. The visit less than a year earlier hadn't worked, so I guess moving to New Orleans wasn't an option. Phone calls ensued, an arrangement was struck. The next day I boarded a plane to Houston, to my sister Kinta and her husband Mike. A few months later we moved to Dallas.

Kinta and Mike had met at the Playboy Club. She soon dropped out of college, married him, and landed a job working as a stewardess for Continental Airlines. Mike bounced from one job to the next, raging that the world had somehow served him poorly: traveling salesman, filling station owner, carting newspapers around in the early morning, jobs that never lasted very long. Abandoned by his father, Mike had changed his last name when he turned eighteen, choosing a popular writer of spy novels whose suave, solitary hero had the best girls and the fastest cars and was always in complete command.

I needed a roof over my head. Anything was preferable to ending up in foster care, or worse. Dad would redirect the child support check to Texas, and I would work to help pay my way. I was expected to leave just as soon as I turned eighteen, and I was fine with that.

Dad jetted back to Los Angeles. I wouldn't hear from him again for nearly three years.

———ᴍ———

The late teens can be a particularly fragile period. This was certainly the case for me, as it is for many Americans. Usually things work out in the end, though as parents we often worry ourselves sick watching our children stepping into adulthood. But some really do go astray. Psychosis, for example, very often first appears in adolescence, especially schizophrenia, which most scientists believe is tied to abnormal levels of neurotransmitters inside the brain. Over 13,000 teenagers die each year in the United States, the vast majority of them violently: car crashes, murder, unintended injuries from reckless behavior, and especially suicide. Sometimes these tragedies are simply the consequence of one bad decision, or the horrible chance of being in the wrong place.

Some kids seem unable to create those stories of the self necessary for adulthood and the separation from childhood, stories that somehow help create the inner compass needed for discovering their future. They struggle finding out who they are. Adolescents with traumatic childhoods are especially vulnerable. Psychologists working with adolescents theorize that our childhood pasts—particularly our early attachments to parents—become an important if not always conscious part of the way we begin conjugating our lives. Somehow the past insinuates itself into the complicated work of becoming an adult. The developmental failures and weaknesses of childhood return, in many instances leading to mental anguish, even breakdown.

The psychological transformations of adolescence are related to changes within the brain, although the precise relationship is the subject of ongoing research and debate among psychologists and neuroscientists. The adolescent brain undergoes terrific alteration; it may well be that these changes are somehow tied to the onset of some mental diseases. Total cerebral matter peaks in the years between about ten and twenty, usually a bit later for boys than girls. Certain neurotransmitters flood the organ. Dopamine pathways, for example, spread from the middle of our brains toward the front of our heads. Studies have identified dopamine as important to brain functions relating to memory and mood. The chemical may also play a role in neural plasticity. Neural connections flourish within the adolescent brain. Dendritic forests appear in critical areas, creating the potential for more developed forms of consciousness and higher level reasoning. The brain also prunes redundant connections, while increased myelination enhances neural transmission. The frontal lobes mature. More generally, the brain attains a greater and more permanent and refined level of organization. Our brains become wired for adulthood.

New research suggests that the development and distribution of dopamine neurons likely is tied to the evolution of human cognition. A 2008 study showed that chimpanzees and humans, but not other apes, have certain dopamine neurons in select cortical areas important to cognition that also indicate plasticity. Scientists reconstructing the genetic history of our species have identified important changes in the periods around 300,000 to 500,000 and 40,000 to 50,000 years ago, eras associated with hominid and human migration. The more recent dates are especially intriguing, since this is the time anthropologists have identified with a revolution in human consciousness, what some have called the "big bang" or the "great leap forward" in our species history. We began burying our dead, fashioning jewelry, telling stories in paint laid upon the walls of caves.

Psychologists refer to the neurological changes inside the adolescent brain, particularly those unfolding in the prefrontal cortex, as the development of "cognitive complexity." We can see this in the endless conversations adolescents have with their peers, in the constant pronouncements followed by indecision, and in the bumptious relations with adults. What begins emerging are narrative structures or "scripts" whereby abstract thinking allows the adolescent to develop not only a sense of self, but a sense that in important ways they "own" that self. This sense of ownership or agency, and its attendant declarations of independence and life in the present and future tenses, is central to the development of adulthood.

None of this is straightforward. Separating from the past, saying good-bye to childhood, also means some sort of engagement with it. Adolescence is a mess, a marvelous, often terrifyingly confusing mess. Identity is pretty much a work in progress. During adolescence the question of who one *really* is can seem remarkably unclear and subject to seemingly endless revision. Neuroscientists used to explain this mess by arguing that the ongoing growth of the prefrontal cortex meant that teenagers were not always capable of rational decision-making. At the anatomical level, however, the situation is far more complicated. There are, for example, important changes unfolding within the limbic system, which plays such a vital role in emotion and memory. One result of this is an increase in what neuroscientists describe as "emotional reactivity," and particularly an attention to danger. Studies show that the amygdalae seem to be especially active in adolescents.

In other words, various areas of the brain are maturing, but not always at the same pace, and all of it in some sort of relation to the rest of our bodies and the surrounding world. This discordant symphony of neurological development may explain why teens sometimes engage in risky behavior even if they know better. This usually exasperates parents, who naturally worry about their child's safety. Riskiness, however, may be absolutely important to becoming an adult and likely has an evolutionary component. Teens are, as nearly fully formed animals, descendants of those willing to take a risk to secure a source of food or desired mate. It may well be that we can only become autonomous selves by taking chances.

—◊◊◊—

My recollection of the eleven months in Texas feels as if it is a past that is not really my own, or that in some sort of hiatus I inhabited a disavowal, a life slightly removed from my self. Memories appear like Polaroid snapshots you discover at an estate sale drifting at the bottom of some box with the assorted bits and pieces of someone's life. Nothing seems like me. A scrawny boy with a shag of dark hair pumps gas, changes oil, and plugs holes in tires at a Phillips 66 station in Houston's sweltering heat. At the back of Spencer's Gifts in a Dallas shopping mall stands a young man selling fiber-optic lamps and Day-Glo posters. He watches the lamps shift from one color to the next and then back again, punches his timecard, goes home. A student sits on a folding chair at assembly, speaking to no one as the cheerleaders in their pleated skirts bounce down the aisle.

Simple chronology determines that I graduated early, a decision I think I made myself. School in Tunisia, followed by California and Maryland, had accelerated my education. Texas standards were low. I just needed a few more classes. I have no a copy of my diploma, don't even remember the name of the school. I just wanted out.

The months unfurled. My father's legal commitment to provide child support ended once I turned eighteen. I would be on my own. I was hopelessly lost. Kinta mentioned Dallas Community College; others simply shrugged. I have the faint recollection of saying that I wanted to return to Maryland, where I could attend one of its public institutions as a state resident. Although we had not spoken, Susan had filled in the forms for the University of Maryland, College

Park. And she had done something more, submitting my name for a state scholarship that would cover the costs of tuition. Estranged, she had nonetheless come to my rescue once again. I was going to college.

EIGHT

—w—

LESSONS

I SAT AT A LONG WOODEN TABLE, HUNCHED OVER BOXES of records stored in an old Edwardian building that leaked every time a storm rushed ashore. Drops of water fell from the domed ceiling into metal trash bins that dutiful attendants had placed around the reading room. One drop turned to three or four, each catching the light on its way down. The metallic pinging sounds turned sonorous, mournful even, as though the archives were weeping.

Outside people spoke of revolution, chaos, violence, civil war. In the closing months of 1984, South African troops had gone into the townships to restore order. Soon the government declared a national state of emergency. There were massacres, mass arrests, disappearances, plumes of smoke rising from South Africa's war zones. The woman who lived in the apartment above me taught at the university and in the evenings helped plant bombs and distributed AK-47s to her comrades in the townships.

In the evenings I attended political meetings and during

the weekends went to the funerals of people who died by government violence. But each morning I arrived in the archives with sharpened pencils and a stack of blank note-cards, possessed by some need, even a feeling of obligation, to transcribe words penned centuries ago, sneaking in early until frustrated archivists bolted the door shut. I pored over the remains of the past believing I might grasp something that always, it seemed, lay just out of reach, the memories and traumatic experiences that might reside in words never intended to be housed in an archive or read by others.

The documents seemed endless: the correspondence of colonial officials stationed on the empire's far-flung frontier, the proceedings of thousands of Africans brought before the Law, the pleadings of people made strangers in the land of their birth, the deaths of tens of thousands, the journals, diaries and letters of the dead. Sometimes the writing was nearly indecipherable. Clever merchants diluted their stores of ink. Words disappeared into the paper, faded beyond recognition. Silence and loss lived amidst the millions and millions of pages and billions of words, something always missing, some persistent emptiness. Dirt and dust spilled onto my hands and drifted into my lungs. Documents occasionally fell apart before me, as if time had eroded the past beyond recognition. A letter might end abruptly with a tear or with paper turned into charcoal that left one's fingertips smudged black, its author unknown. For days my hands smelled of a fire a century or more ago.

I took my notes. There were lonely, feverish nights surrounded by fragments of the past mistakenly bequeathed to the present. What might these bits of paper hold? What secrets did they keep? What remained forever lost? I wanted to be able to explain the horrors that unfolded and were still unfolding at Africa's southern tip. I wanted to give the dead another chance, another life even. In my Cape Town apartment I began assembling my notes into some logic of organization and argument, trying to figure out why things happened the way they did. Words followed, hypotheses were discovered, discarded, refined, until a dissertation emerged, and a degree, and a career teaching history.

—⚯—

"Come to New Orleans," Sabrina had told me six years earlier. I had just turned eighteen. The scholarship wouldn't put a roof over my head, or food, not even pay for books. I hadn't filed any papers for the federal government's Guaranteed Student Loan program. I just figured I would hitch up to Maryland, find a job, start classes, and somehow everything would work out. I hadn't a clue what I was doing. New Orleans's restaurant industry offered a convenient way for young people to make money busing or waiting tables, or laboring in the kitchen. Sabrina earned enough money to take a few classes at the University of New Orleans. I would need to do the same thing.

I worked nights at Commander's Palace in the Garden District, carrying heavy trays of food to well-heeled tourists and the New Orleans elite, then at a newly opened restaurant in Carrollton owned by two eccentric women, one of whom drank Chablis Cassis all day and danced across the dining room. During the lunch hour I waited tables. I worked the kitchen at night, helping prepare dishes like redfish meunière and shrimp creole, good classic New Orleans fare. Susan's Tunisian parties came in handy—serve left, remove right, attend to the women first, silent, nondescript motions as if you weren't even there. I learned how to chop properly, sauté, prepare complicated reductions.

The next summer I cooked and cleaned toilets on an oil rig twenty-six miles off the coast. I washed dishes watching gigantic fires draw lines of light against the Gulf's still waters. The other workers mostly came from the gentle arc of land stretching from Houston to Florida's panhandle. A few had drifted from up north, fleeing whatever troubled them, hanging out in the Big Easy living too hard before heading to one of the rigs. Good money was easily had if you were willing to work hard and long for a week or two with your body twitching and your eyes hungry for another line of speed or coke. You'd be back soon with money stuffed in your pocket, and the crazy would start all over again in some cheap dive off Canal, scoring in some back alley, hanging out with strippers and barflies, getting wasted just as soon as you woke late in the afternoon. You'd burn through all your money, and you'd be standing there again, hung over, a stubble across your face, drawing on a Camel or Marlboro between gulps of coffee from large Styrofoam cups, waiting for the boat's diesel engines to pull you away.

Work kept me out of my mother's apartment. Mom had tried killing herself again shortly after having to leave Chestnut Street. Benji, my sister Sabrina's boyfriend, had found her by pure accident and carried her off to the Emergency Room. Mom tried tending to herself by putting a few flowers into the backyard and by painting geraniums she had potted in coffee cans. She checked out mysteries from the local library and stared at the nightly news. By the weekend, when it was time to buy milk, eggs, coffee, and cans of Campbell's soup and Saltine crackers, it was impossible to avoid the quart bottles of beer cooling themselves behind sheets of glass dripping with condensation. Mom would buy a few, drink until she passed out, and stay in bed until the early afternoon. On Monday she stepped up to the streetcar on Carrollton, got her transfer ticket, and headed to work, gazing at the lead-glass doors of the stately homes along Saint Charles Avenue.

—⁓—

The University of Maryland campus seemed foreign, with its clean brick buildings and verdant lawns, its students carrying heavy textbooks or lolling about the quad. I remember meeting with a graduate student in some noisy building, poring over a thick book of course descriptions printed on tissue-thin paper. We sat at a folding table. He had a round, gentle face. There were charts to complete, requirements to take—science, foreign language, writing composition—but room to experiment. The brief mention that I had lived in Africa led him to courses in international relations and African history. A schedule began emerging from my scarcely audible statement of interests.

In the late 1970s, full-time faculty still taught most of the classes at major public institutions. My biggest class was introductory biology, and even that course was not more than sixty students. At an institution of some thirty thousand, one could easily take lecture classes of thirty or forty students, and seminars as small as six or seven. I enrolled in the usual suspects: biology, botany, French, but I also took courses in international law and development, intellectual history, and the history of Africa. The library seemed an infinitely large refuge, one floor after another of books stacked on metal shelves offering the solitude of reading at a table next to windows coated with a thin grime. I purchased as few books as possible, which meant that I lived in Course Reserves, presenting my ID and walking away with a text with a band across it commanding I return it in two hours.

My scholarship stipulated that I had to maintain a certain grade point average. I lived with the certainty that I would fail, that this whole college idea was a fantasy. I might as well hitch back to New Orleans where I belonged and begin a life as a cook or learn some other manual trade, stoned most of the day. This was the truth of my existence, everything else an adolescent phantasm. I pulled myself from bed at six in the morning, poured grounds into a Mr. Coffee, walked to the library, struggled to understand the words drifting across the pages of books and articles: the concept of just war, the Geneva Convention, the law of the sea, the role of organelles, the spread of Bantu languages across the African continent, a bewildering torrent of concepts and facts. I didn't know what was important. I felt hopelessly lost.

I sat hunched over the exam booklet writing about the state in Africa, trying to recall the arguments of a slim volume by a Cambridge don on whether or not the concept of feudalism had any purchase on the study of Africa's past. There were not more than four or five of us sitting at the table.

The professor tapped my shoulder, leaned over. "I would like to speak to you in the hallway after the exam."

She was a tall, elegant woman in her forties who worked on West Africa. I stood beneath the fluorescent lights leaning against the painted cinder-block walls in the basement of Francis Scott Key Hall.

"I want to encourage you to continue in your studies of Africa," she said. "I'll see you in the fall."

A few days later I filled a backpack stuffed with clothes and a few books and walked to Highway 95, thumbing one ride after another back to New Orleans.

—⁂—

In college we begin making real the scripts we began creating about our selves in adolescence. We imagine a future, not only in our heads but in the complex community and inherited culture that is humanity. The maturing brain's cognitive complexity helps make this possible. Recent research suggests that our memory systems fully develop only in early adulthood, allowing us, for example, to trace memories to specific origins. We can embark on the journey of considering the self and *its* past, which means that we begin facing all the complications and contingencies of who we have become, our life's journey. We can bring a kind of history to experience. This appreciation or awareness of the past paradoxically helps us separate from it. We say good-bye to childhood, which also usually means saying good-bye to our parents.

The challenge is not so much a separation from childhood as it is an integration of the past and our imagined future. What purchase does this past have? Humans have a universal need to form close bonds, beginning in infancy with attachments to caregivers. At times these attachments are less than optimal; there is no such thing as the perfect parent. Sometimes they are terrible, as in my early relationship with my mother. I likely folded her despair into my emergent self, where the past was less consciously remembered than felt and the present seemed forever unreal.

In early adulthood's gloaming I journeyed between two belongings, one exerting some powerful gravitational pull toward a New Orleans childhood I could not summon to memory, the other indeterminate, indistinct. College, the comfortable waltz of students across the quad and Frisbees curling in the air, seemed like a fiction. I couldn't understand what I was reading. Words floated past in some indecipherable whir. I went to class, took my notes, but the world kept slipping away. And I didn't have the money to make ends meet.

I often woke in the middle of the night covered in sweat. I would dream of running endlessly through suburban subdivisions. Or I would be in New Orleans roaming the French Quarter or sitting on the couch in my mother's derelict apartment, legions of roaches marching across the floor. I wandered through my sophomore year dreadfully lost. The world seemed unbearably heavy. Everything that had unfolded in the present simply disappeared, returning me to my New Orleans childhood, to a house and a neighborhood, to a past that felt real but which I could not describe or summon by the simplest of words that might help distinguish the now from the past.

A doctor at the clinic prescribed antidepressants. I walked into woods near campus, sat down with my backpack next to me stuffed with books, and gulped down the entire vial.

Depression steals time, erasing the horizons of our inner being upon which we direct our lives. The timing of my crisis was not unusual. In late adolescence our past begins meeting a present that is ours to shape, often for the first time. Will we become our stories, however roughly conjured they may be? Or will we somehow yield to an earlier era when we may have been the subjects of another's life? Adolescent breakdowns occur for all sorts of reasons. Problems that had emerged much earlier in life may suddenly begin surfacing. In some instances they emerge from our inability to find our mental selves and to reflect on why people do the things they do, what unfolds in their minds. Something remains fundamentally and persistently unresolved. Unable to locate our selves, we assume the mind of the other even as we search desperately for some sort of autonomous existence. In crisis, this untenable conflict ruptures.

I discovered these issues not only in a therapist's office but in my studies. Throughout that troubled year I found myself pulled to the work of social historians who were committed to uncovering the lives of workers, peasants, and women—common folk. Theirs was a work of recuperation. At the center of this scholarship lay a vexing issue, what scholars called the problem of "structure" and "agency." I wrote down a quote from Karl Marx, which I had in my backpack that day in the woods: "Men make their own history, but they do not make it just as they please; they do not make it under circumstances chosen by themselves, but under circumstances directly encountered, given and transmitted from the past. The tradition of all the dead generations weighs like a nightmare on the brain of the living."

What structures determined my world? What agency?

—ɯ—

A piece of mail arrived that spring that had made a circuitous journey. Beginning with a trip from some large Washington, DC, building to my sister's house in Dallas, it was redirected to my mother's Zimpel Street address. It must have sat on the couch or on a table for some time until Mom scratched out one address and put on another. I am not sure how she found where I lived, whether she called me or someone else, but somehow the letter traveled back north to Maryland to just a few miles from where it began.

It was one of those official letters with the name of the bureaucratic agency neatly printed in the upper-left corner and a cellophane window, from the Social Security Administration. From 1965 through the mid-1980s, the federal government supported college students whose parents had retired or were deceased. This was the golden age of American higher education and social welfare policies that allowed millions of students like me to attend college. Institutions such as the University of Maryland grew enormously. The Social Security Administration provided an additional cushion, and encouragement, for kids to go to school and, crucially, to stay in.

The letter was simple. As long as I remained in school, each month the government would issue a check, in my case $200. Dad had retired at the end of 1978, halfway through my first year in college. I had accrued a number of months of income. With a few forms completed, the money would be mine.

A second letter arrived, this time from my father telling me that the money belonged to him. I was to arrange for the check to be sent to California. Dad would then send some of it back, whatever he thought I deserved.

I didn't write back.

Then came a phone call one weekend; I think it was a Sunday. I was living in a student house north of campus, a fifteen- or twenty-minute walk to class. I can't remember the exact words. I know he said "son" more than once. This meant he was serious, as if a word indicating our biological connection automatically imparted gravity to what followed.

"Son, I'm retired now. I've been sending your mother money from the beginning, son. Susan too, when you lived with her. That money belongs to me—I reckon I've worked all my life for it. You make sure the check gets sent to California. I'll take care of it for you. I'll send you some money like I've been doing all your life."

"No, Dad. I'm not going to do it."

I don't remember ever speaking up to my father like this. I had read the bureaucratic language. The money was meant for me, for the children of retirees. My father had rarely been around, and here he was calling me from California, reaching out because the government was cutting me a $200 check.

"It's my money," I told him. "It's supposed to go to me, not you."

Dad said something else, insisted the money was his. I could tell he was furious. I put the phone down. He stopped calling.

With the tuition scholarship and the government check, college seemed possible again. But I would still have to work to make ends meet. I found a job at the Sunrise Café, a diner along a jumbled campus strip of bicycle shops, pharmacies, bars, and cheap places to eat. I would stay up all night on Thursday finishing my reading, typing up my scrawled, barely legible class notes. I worked the Friday night shift, then all day Saturday into Sunday morning, twenty hours washing dishes and making sandwiches and omelets. This way I could spend the rest of the week reading in the library.

I didn't return to New Orleans at the end of the school term, nor the next summer. I rarely called or wrote. Years passed by.

I know now that in those months after I woke shivering in a Maryland woods, some part of me realized that to live, to demand ownership over my life, required a kind of disavowal. One can exist in an alien world where the self remains tied to a past over which one has little or no say, in my case weaving my mother's despair into my inner being. Or one can begin the awful, lonely work of claiming a future. For many of us with troubled pasts and suffering minds, disavowal becomes a necessary condition of survival, not so much repudiating our history as renouncing its continued haunting place in our present-day lives.

That spring I filled in the forms declaring a major. I was going to become a historian.

—✺—

I am drawn to impossible histories.

I realize now that I became a historian because I wanted to write into the wounds of the past. My professional life is bound to a childhood of neglect and dislocation and, most profoundly, a difficulty remembering my own past. If I could not have my own memories, at least I would try to preserve others. I turned to South Africa because at the time it was the most brutal place on Earth. In some ineffable way, I transferred an inner narrative onto a historical landscape that seemed far away from New Orleans, trying to figure out what happened, what went wrong, and why.

In the late twentieth century, South Africa had the unenviable distinction of being the world's anathema, what Nazi Germany had represented to an earlier generation: intolerance, brutality, an entire political order manufactured to persecute a people on the basis of race. I wanted to write a history of this inhumanity, determine what made this traumatizing present possible by a determined wandering through the detritus of a bygone era. I thought, naively I know, that the archive as a house of memory might allow a return to some inaugural moment, the point where the damage was done.

—⚬—

We write our own histories, but not according to conditions of our choosing. Constrained by what is or isn't in the record, the historian brings to the past the perplexities of his or her age, sometimes consciously, often unknowingly. We seem to work with what's there—the evidence—but it's what's absent that often drives our longing, our worried rowing toward worlds already slipping away.

Historians believe our truth lies not in metaphysical reflection but in the hope that we will discover something in the details of another's life. All the while we are drawn to traumatic pasts, to the silences produced by terror and brutality, the infinite frailties of existence. We tell stories that move toward some end in time knowing that we are surrounded by fragments and absences, by all that's not there, that's forever lost. History, like art, may be the preserver of memory, but like the amnesiac it also remains removed from human experience. We yearn for what is missing. The past erodes and dissolves, even disappears entirely, at times as fragile as the memories we keep within us. And sometimes just as powerful. The historian works with bits and pieces, sometimes bare traces, as if we were survivors walking through history's wreckage trying to make sense of it all, surrounded by the remains, certainly, but also by silence and forgetting. What's there in the human record is often as important as what is already gone, the thing for which we stand ever longing.

—⁓—

Perhaps history and the new science of memory are not dissimilar. Scientists once believed that the brain was more or less inflexible, its structures discretely organized. Memory entailed the recovery of data stored inside us, our heads a kind of filing cabinet of past experience. Problems emerged only in the processes of retrieval and interpretation, as if a piece of data had been incorrectly filed and temporarily lost, or we misunderstood what we remembered.

Memory's retrieval and interpretation became a central goal of psychoanalysis and other forms of "talk" therapy. The mind repressed traumatic or overwhelming experiences, so the goal of therapy was to make repressed memories visible. Each revisiting of childhood injury, each awakening of memory under the analyst's wise supervision, would leave the adult with a better understanding of their past and its location in the present. Patients would leave therapy with a new history of their selves, and with the understanding that we have control over what once seemed inevitable. Trauma would no longer seem timeless. The past, finally, would become past.

Only the most orthodox psychoanalysts now believe in the "repressive hypothesis." Memory is not a physical "thing" we can lay hold of. Memory amends itself over time, part of the endless little additions and revisions that make a life. The brain is ceaselessly changing in ways that radically differ from any other part of our body. To a certain extent, the external world, that is, culture, shapes neural development. What happens around us has an impact on how the brain encodes subsequent experiences as "memory." The past that creates us in turn shapes what we make of the world and how we live in it. Memory may be as delicate as a wisp of smoke and as resolute as fired clay. In it resides our past and the past of others.

The archive of the self, memory is the way the human brain makes sense of experience. Our brains are organized for telling stories of memory. We have narrative brains, prepared for producing and communicating history. Telling the past seems unique to our species. The development of neural structures inside the brain related to memory, such as the amygdalae and the hippocampi, may be related to the fact that mammals bear live young that they tend and defend, and that they typically form social communities. The origin of these bonds goes back to the emergence of the mammalian brain over sixty-five million years ago. History, that is, stories about the past, serves an evolutionary role by translating experience into knowledge that helps us, and others, navigate a world and explain our location in it. We are fundamentally historical beings.

The ways memory is formed, arrives, persists, fades away, disappears forever, the stories we tell others, and ourselves—all of this is history. Even involuntary memories, those that emerge willy-nilly from external stimuli, seem to be the brain's way of reminding us that we don't only live in the present. The past remains forever unfinished because in powerful ways it is inside us and all around us in human culture. Its presence compels us to tell stories, to tell them again and again, and to hope that someone will listen. This is one of the burdens of being human, remembering—insisting—that stories remain ineluctably open. We continue living in the telling.

—⟩⟩⟩—

I still cannot remember most of my New Orleans childhood, despite all the time and effort I've spent trying to reconstruct it. The external world simply overwhelmed my young brain's ability to put things into some sort of order. The issue is not one of repressed memory, or at least not only this. My brain simply shut down, adopting a cognitive strategy of not remembering, a primitive way of declaring "Enough already!" Certain experiences disrupted the complex interaction of the brain's different memory systems. Some sort of shutting down—call it repression, inhibitory selection, forgetting even—helped mute the besieging external stimuli. All of this conspired to create problems with autobiographical memory.

Familial dissolution also played its part. Kinship groups have been the single most important site of historical production, whether it's about how to survive or the wisdom held in the lives of our ancestors. The destruction of these groups by violence, poverty, or divorce breaks not only the transmission of knowledge but the very ability to tell stories of the past

There is nothing much I can do about this. This is what it is and what can never be. I grieve for what I cannot remember. It's a peculiar mourning. Lost childhood remains stubbornly present, its absence an abiding life.

I have dug into a family past that was lost to me. I conducted interviews, pored over records, sat in bars, walked up and down streets, read books in the library, traveled to distant archives, bowed down before the therapist's couch. I followed the scholar's method of verifying information through independent confirmation, a kind of triangulation historians use to guide their way toward some sort of meaning amidst human despair and time's unremitting erasure. I could glimpse someone else's past in the most mundane of accounts and, most powerfully, in the coincidences that lay among the shards. This is important. History can be made in the most unexpected ways by the jostling together of bits and pieces that were never meant to meet.

All this documentary evidence of someone else's experience, even my own, remains just that: evidence. It can never become memory. There is no return from the nothing-ness of forgetting. History is my artificial limb connecting the present to some distant past that is forever missing.

But there is something different now. It's not memory but still powerful: the knowledge that helps fill in the blanks spaces where a child once walked all those lost years ago.

—ɷ—

All of this raises the vexing issue of survival that scientists have tried to explain by looking at issues such as genetic pre-disposition, timing, and resiliency. Why do some people seem able to "make it," even thrive, while for others trauma leaves them destroyed? There are no objective criteria for trauma. People process external stimuli differently. Some have experienced things that seem impossible to endure—rape, war, genocide—yet are able to put together decent enough lives. In some instances, trauma simply isn't trau-matic. For others the situation is vastly different. They are forever haunted by their past.

I was not abused in the usual sense of the word, the stuff that makes it to the nightly news. I neither experienced a single instance nor multiply repeated instances of things such as beatings, sexual assaults, and so on. I was, rather, systematically neglected over the course of my early life. This was traumatic, but in radically different ways—the dif-ference between a mother's constant abusive drunkenness and witnessing a murder. It was exceedingly common, the garden-variety trauma that happens to millions and millions of children every day.

It is also typical for people who have a chronically difficult time remembering their childhood past. I know now that in addition to memory problems, externally the child may create various attempts to cope by trying to block off or retreat from the world, often at the very time they are experiencing something traumatic. Later in life they may have the tendency to reenact in their daily lives a past they don't quite remember, let alone understand, but which still takes possession of the present. Experiences of what happened an hour ago, the embrace of a loved one, the color of flowers tended all spring, vanish. Their instinctual core dissolves, even the very will to stay alive surrounded by so many kindred ghosts.

Children of broken families often bring upon themselves the burden of trying to stitch together what is irreparable, and the guilt for having failed at that. Sometimes they seek impossible relationships, as if they were recreating in adult life the heroic if unbearable work of childhood. Or they will not allow themselves to let go of life patterns that are destructive. It seems hardly possible for them to walk away, to be able to say they have the power to make their lives. The story of the self struggling and failing against an overwhelming past is just that, a story. It is true. But it is also not inevitable.

An inordinately large number of my extended family have taken their lives or tried to: jumping off bridges, overdosing, pulling off a Louisiana side road and putting a .357 Magnum to their head. My siblings live in despair, with broken marriages, depression, abusive relationships, and substance abuse. Marie spent her entire adult life trying not to be like our mother only to become just like her, an addict, falling down steps stone drunk, crushed by the past. Kinta dreamed of a prince but instead she's stuck in a four-decade-long abusive marriage. She's packed her bags and driven away countless times, but somehow she always turns around and comes back home. Sabrina is married to an alcoholic and keeps herself in a near-permanent marijuana stupor. Susan has gone through multiple marriages and is beset by loneliness. My brother lives in dire poverty, his life a ruin, a girlfriend deported back to Central America following her arrest for stealing copper tubing from AC units.

—⁓—

Who saved me—my grandmother, siblings, teachers? Did I save myself? Or was I just plain lucky? Did history's contingencies somehow conspire to allow me to get through? Perhaps I have been saved by my amnesia, the absence of memory as some primitive defense? Forgetfulness entails casting memory into oblivion. No wonder amnesia shares a root with the word "amnesty," the forgetting of sins, the letting go of too much painful history.

It is impossible to disentangle all these questions. I was one of the pieces of a wrecked marriage, conceived as a mistake and brought into a world that was breaking apart from the titanic forces of alcoholism, mental disease, divorce, and poverty. Unlike my siblings, however, I did not face all the pressures of trying to put the pieces back together again. I got to escape, if only for a few precious years, into the stability of the middle class and especially into the attentions of a few wonderful teachers who by their instruction showed me a way out. I was not the one in the car pulling Mom from dingy bars, nor signing the papers for her release from the mental hospital. I was not the one who stitched together Christmas stockings from pieces of red and white felt, insisting that there was a holiday to be celebrated, no matter what, forcing Mom into the kitchen to cook her exquisite oyster dressing before she hit the bottle and passed out by early afternoon. I was not the one who had to break down the bathroom door. I was not the one who remembered.

Neglect wreaked its damage. The terrors that have befallen others have called to me. I still occasionally tumble into my second-person self. I also perfected a guilt over not being able to fix everything that went wrong, particularly my mother's poverty and despair. In other instances, however, neglect offered a vantage point, a kind of perspective. Perhaps I wondered about what was happening, the meaning of it all, without necessarily having to do anything beyond surviving, the solitude of living on oblivion's edge.

We all grapple with our pasts, trying to fathom the story of our present—even with those parts of it that remain lost to us. Sometimes it's the empty spaces that call most powerfully, the source of our despair and wonderment, the knife's jagged edge but also the point where the work of healing begins. Like our brains, history is constantly changing, subject to the endless revisions that compose who we are—our pasts and our future.

EPILOGUE

—⁓—

I KEEP RETURNING TO NEW ORLEANS. THERE IS research to conduct, and there are professional conferences to attend. But most of my time seems to be taken up with the dead and the dying. It's one of the peculiarities of middle age, watching our parents age and pass away, the obligations to care for them in their final years, and, beyond, their memorialization. There are family plots on both sides of the family: a nice Salvant vault in the St. Louis Cemetery No. 3 just off the Esplanade, and a more modest plot in the Masonic Cemetery for the Crais family.

I began this pilgrimage of memorialization some two decades ago when I buried my father. While I'd seen him in the years following our argument over the Social Security check—a sister's marriage, then a niece's—we never really spoke to each other. In the spring of 1990 I was out in Los Angeles interviewing for a job. I had a free day. I thought about giving him a call, renting a car for the hour-long trip to the retirement community where he lived with Letha, but I decided not to. He died a few weeks later. Dad arrived in

New Orleans a few months after that via UPS in a simple, nondescript square box with lots of tape, dutifully packed by his widow and third wife. Sabrina put Dad's ashes on the mantle but didn't make arrangements for his burial. Sabrina hadn't called the cemetery, hadn't filled in the forms, hadn't told them that on this particular Saturday morning in the month of June we were going to lay to rest the remains of our father. Kinta flew in from Texas, my sister Marie and her family piled into the car and drove from Florida. My wife and I headed down from Ohio. New dresses were purchased at the mall, along with lipstick, perfume, pantyhose. I polished my shoes. We went to the ATM to withdraw cash, so that we could all feel responsible when it came to pay a restaurant bill. We were here, in New Orleans, ready for a funeral. Sabrina's father-in-law was a minister and had agreed to say a few words. We steeled ourselves. We wept in private and now were ready to cry standing alongside one another—family.

And Sabrina had spaced out. The authorities had not been notified. The caretaker hadn't a clue. We couldn't have a funeral, and we couldn't bury Dad, at least legally. So Sabrina slipped the caretaker some money. He told her to arrive really early in the morning, before visitors began arriving to pay their respects. So we woke up and headed down the street to the cemetery. Kinta was holding the golden box of ashes. I had a shovel balanced across my back. Sabrina was talking. A few people saw us and, I suppose, innocently concluded that we were going to do some tidying up, weeding, cleaning, making sure the family plot was up to snuff.

I had the shovel resting on my shoulder like a miner going to work. I would do the digging, they had decided. I thought about death and burials as we approached the cemetery. There are ten family members buried in a plot about the size of a couple of sheets of plywood, going back to 1939 when my grandfather died in his mid-fifties. This was my first visit. It's a small cemetery, with most people housed in little marble temples. Since most of the city is below sea level, and the water table is just a few feet below ground, New Orleanians have long buried their dead in the fresh air. Families had their own mausoleums, with shelves on the inside to support the coffins. My grandmother, a good Catholic to the very end, is in one alongside other family members. I liked to think of them as summer houses for the dead, places where the departed could go for an eternal vacation with other family members, until I start remembering just how fucked-up our family is—the divorces, affairs, alcoholism, abuse, madness, suicides—at which point the idea of being stuck in the dark began to seem disturbing, a kind of purgatorial family gathering.

Sabrina wasn't sure where in the cemetery to go. She hadn't gotten a map from the caretaker, who was nowhere to be found and, I suspect, had burned through the money and was now sleeping off a generous hangover. Here we were, with a shovel and a box of ashes. And we were lost. The cemetery had a few winding lanes, but there were graves clustered behind, so we wandered about for a good half-hour before we located the right plot.

But the caretaker hadn't indicated where Dad was supposed to go. I expected a flag or some sort of marker: "Dig here." There was nothing but a patch of grass.

"Sabrina, where am I supposed to dig? You didn't ask the guy where?"

"Just dig a hole." The issue didn't seem to faze my sister.

"Whatayamean? What if there is, well, like someone else already there," I said, pointing indiscriminately. It was one thing burying Dad on our own, but something else entirely digging up someone else. I had this vision of the plot turning to little piles of dirt, of boxes upended, some corroded and spilling their ashes onto the soil, and my sister Kinta still holding the box of ashes, then being arrested for conducting an illegal burial or, worse, defiling a cemetery, grave robbing. I started pressing my shoe into the soil, hoping to find some vacancy, an empty space below. Soon all three of us were walking around the plot like those treasure hunters on the beach with metal detectors. But all we had was a spade, one sister already high as a kite, and another patiently holding Dad and wondering if we were going to end up returning to the house, defeated. We were going to end up faking the ceremony. It would be a little lie for the three of us to share, like children who had done something wrong and had sworn an oath not to tell their parents or older siblings. We could even send the box of ashes UPS to one another if one of us got tired of the ruse.

"How 'bout here?" I said, somewhat exasperated and pointing randomly. I was becoming paranoid, worrying that the police might arrive and we'd be hauled off to jail. Dad would be impounded. Dad would become evidence of a crime. The funeral would be canceled. Everyone would be angry, really pissed off, that they had made the trip home to New Orleans only to have Dad confiscated by the men in blue and sitting on some shelf downtown along with contraband, rape kits, knives, and guns.

"Yeah, that looks like a great place, yeah, perfect," Sabrina said with absolute confidence as if she had some X-ray vision that had identified the exact open rectangle of space amidst a jungle of bodily remains.

It was one of those spades with a straight edge so that, ideally at least, I could cut neat lines in the grass instead of a bunch of silly arcs. I didn't want to make a mess. I wanted my hole neat and squared like I had seen on television soap operas when widows, lovers, and disjointed men watched a coffin descend neatly into a hole that seemed punched out of the earth.

I did not know how deep to dig. Six inches, a foot, all the way to China? But I managed to scrape the grass from the soil, place the turf to one side, then continue digging. A small pile of black earth formed at the hole's edge.

"I think that's 'bout right. Don't you think so?"

Kinta and Sabrina agreed.

"Okay," I said. Kinta kneeled down and placed the brass box into the hole.

"Yeah. It looks just fine," Sabrina said. "Hurry up. Somebody's gonna see us. We're gonna end up in jail for grave robbing."

I scraped some soil on top, then bent over and returned the grass, patted it down, so that the plot looked like it had been patched.

"What are we going to do about the extra soil?" We couldn't just leave it there, a small black mound like the leftovers of some construction project. I started spreading the soil around, which only made things worse, creating a dark smudge on the plot like a child's finger painting. After a while I gave up, hoping that by the time of the service the dirt would somehow disappear back into the earth, or that it would rain, or something, but that people wouldn't notice. I could just tell people that Dad was buried, and I could point. And they would nod solemnly, and we could all go on with our lives.

We walked back. It was already a boiling hot New Orleans summer day. In the shower I watched the dirt spin away down the drain. I wept. I worried. Children develop that inexplicable responsibility to their parents, even if they barely know them or—perhaps even more so—if their parents have been dreadful. Despite everything, I love my parents. My father wasn't bad. He never beat us, nor was he, like my mother, a drunkard. He never did much of anything. That was the problem. In the second half of his life when I knew him, my father wasn't there. And even in those few years when I lived with him, he was absent—working, falling asleep in front of the television, trying to find another woman who might take him in. And when I was old enough to confront him it was too late. His brain had gone to mush. And even if I had railed against him he wouldn't have said anything.

And still, however distant we were, however much I felt abandoned (and I now knew he had abandoned us in one way or another), I felt I owed him a decent burial. I realized I was still searching for his love a long time after he was gone.

That afternoon we returned to the cemetery for the service. Although they had been divorced for nearly thirty years, my mother arrived with a dozen red roses. She stood at the back, behind me and my sisters and a motley arrangement of nieces and nephews. Reverend Madden said a few words, none of which I remember, because I was still fretting over that scratch in the grass and looking overhead to the thunderheads starting to come in from the Gulf. I knew it was going to rain, and that the rain would wash the soil away and that the grass would slowly suture itself, and I would forget where exactly I had buried Dad.

Reverend Madden asked each of my siblings to come up and place a flower on the plot and, if they wanted, say a few words. I thought of Sabrina's loquaciousness, which my other sisters shared, though none as serious as Sabrina's. Each of them walked to the grave, flower in hand. Once they said a few things the words just started flowing, but as I had experienced inside my head earlier that day, they were all directed not to the living, but to a square metal box near their feet. There was no eulogizing, no disquisition to those of us standing about, about our father's life, as a man, a husband, Dad. They were talking like they always had wanted to. It was as if they had told him to sit in a chair and listen to what they had wanted to say all these years when our father was everywhere except where he was needed.

I guess this is why when it was my turn I just placed a flower on the plot, looked at my work, and turned away. I think I had said everything already. The grass looked pretty good. Everything was going to be okay.

While I was completing this book Susan, my oldest sister, my other mother, died following a decade-long battle with metastatic lung cancer. I called her once or twice a week, and visited whenever I could, especially in the final days. Against the assault of chemicals and the disease's relentless spread, I tried offering what comfort I could. She lay in bed curled into a ball, a mere seventy pounds, the pain stealing the little energy she had left. I would tell her stories about the good times in Tunisia, the smell of rosemary in the garden, the Mediterranean as blue as lapis lazuli, our trip to the Sahara with Pablo, until the morphine carried her away.

I listened to her fears and regrets and to her wish to be buried with our father.

"Yes, of course," I tell her. "Yes, I can do that. I'll bring you home." Home being New Orleans.

Susan tells me exactly how she wants her name spelled. I email family members, make the necessary phone calls. The engraver kindly visits the plot to make sure there is enough room on the granite slab for my sister's name. Sabrina thinks we should sneak into the cemetery, perform a kind of guerrilla burial like we did with Dad. Susan would appreciate the transgression, Sabrina thinks. I am not so sure.

Even Mom has returned. In February 2011 she sustained a terrible accident. Mom and Kinta had gone to Stein Mart for the big clearance sale. Across Florida one store after another had closed down during the Great Recession. Property values collapsed. People walked away from their homes. Unemployment spiraled upward. FOR LEASE and COMMERCIAL SPACE AVAILABLE signs stood in vacant lots. Employees received notice, spent their last days at work taking stores apart. Everything goes, even the bathroom sign.

An employee pointed Mom in the direction of the bathroom. At one end of the store were two identical entrances separated by twenty paces: one for the bathroom, the other a changing room. The store had a problem with theft, particularly in the changing room, so an employee placed a heavy display wall halfway across the entrance, leaning it precariously. From a distance, and in a rush, it looked like a louvered door. With the slightest pull the display wall came tumbling down on her slight frame, breaking a hip, cracking one femur all the way to the knee, and crushing a right arm.

An ambulance brought her to the nearest hospital, where she endured multiple surgeries followed by bleeding in the stomach, heart problems, severe edema, and an opportunistic infection that required expensive antibiotics. One wound would not heal, exposing a piece of titanium holding a limb together. On the bulletin board in her room nurses had posted a DO NOT RESUSCITATE order.

My mother's injuries ended a period of her life lived mostly independently, if negligently and certainly in poverty. Despite her age and declining health, Mom had managed to stay in her small Florida apartment. She lived much as she had for over fifty years: with a flea-ridden cat; roaches scampering across floors and walls; maggots growing from pots of food left on the stove; a stack of bills she felt entitled to neglect. Kinta visited regularly. They went shopping. Mother read and painted, and on Sundays leafed through the advertisements and human-interest stories scattered across the local paper. Sometimes Marie would come by and try cleaning up.

"She wants to be buried with Grandmother, in the Salvant tomb, back in New Orleans. Would you find out? It would mean a lot to Mom, to know she could be with her mother, be back home."

I called the cemetery, discovered who owns the legal rights to the tomb. I talked to a distant relative living in

Metairie, offering genealogical information connecting my mother to Joseph Salvant Jr., her grandfather. Yes, she can be buried with Grandmother when the time comes.

Mom also wanted me to help with the legal morass. Months in hospitals and rehab units generated enormous medical expenses. She relied entirely on Social Security. Medicare covered many of these costs, but not all. In less than a month she expended the entire coverage on rehabilitation. Where she was staying charged nearly $7,000 a month. Medicare initiated liens against Stein Mart. Mother's attorney filed a claim against the company, followed by a lawsuit. Dates were set. Interrogatories demanded. Mom and others would be deposed. I studied theories of liability in personal injury law, and how the legal system determines the value of life and human suffering. Most of the elderly who suffer similar injuries perish within a few months, and Stein Mart's initial strategy was to hope that mother died. She had no economic worth, and death would conveniently get rid of any future costs. But Mom hung on, and Stein Mart agreed to mediation.

The injury and resulting lawsuit brought me back to Florida. I sat with my mother in the hospital, describing my work, asking a few questions when I thought she was up to it. She kept her spirits up, and even a sense of humor. "Son, thank you for your visit," she once said. "You've greatly enriched my death."

The insurance settlement provided my mother with more money than she has ever had, though the funds have been swiftly depleted by the costs of living in an assisted care facility. Mom grew tired of the Jacksonville Sunrise. She felt isolated and lonely. Most of the residents came from the

Northeast, solid middle- and upper-middle-class folk, the sorts of people Mom had catered to when she sold cosmetics decades ago. She wanted to go home. She wanted to go back to New Orleans. For just over a year, she moved in with my sister Sabrina, not long after they had completed rebuilding the home that had been destroyed in Hurricane Katrina. At the beginning of 2013 this proved too stressful for my sister. Another flurry of phone calls ensued, discussions of what to do, where Mom might go, how to explain that Sabrina just couldn't cope any longer.

I was deputized to explain to Mom that she had to leave. We sat in my sister's kitchen around a fruit tart I had brought from a fancy French bakery. Her eyes no longer seemed as forbidding. Mom seemed forlorn, scared even, and certainly abandoned as I explained why she had to leave, and that I had found a new place on Magazine Street, not more than a stone's throw from where she lived much of her life.

She seems happy now, or at least happy enough for a woman in her nineties and in rapidly failing health. There is the *Times-Picayune* to read for entertainment, though now reduced to just twice a week, and a flock of parakeets that have been set free in the atrium and nest among the plastic trees. The residents are almost exclusively longtime New Orleanians, with accents that melodically slur one word into the next and stories of Mardi Gras and the city's never-ending commotions. Mom has her nightly dinner with a friend, an elderly woman from Uptown.

In the last few months her condition has taken a turn for the worse. Mom can barely move her walker ten feet. She complains of pain in her broken hip. Mom is wasting away from congestive heart failure. She's also not eating; in just a few weeks Mom sheds nearly twenty pounds. She spends almost all of her time in bed. And just as Susan dies from cancer I learn that during a recent trip to the hospital the doctors discovered a tumor in one breast and suspicious spots on her lungs. I phone my sisters. Marie is convinced Mom is about to die any second. Sabrina thinks she has a few more months. I don't know, but I make sure the nurse has a copy of her living will and the "do not resuscitate" order.

I call once a week, sometimes twice. Lately I've had to pass on the sad news of Susan's illness and death. "Mom, Susan's really, really ill. I don't know how much more time she has." Then, "Mom, Susan died. She was at home," and I explain to her the services, how we are all set to converge on New Orleans for the memorial.

"I am so sorry," she says, as if she were offering an apology.

Her mind is slipping away. A nurse picks up the phone the next time I call. "Your mother's dying," the nurse tells me. She needs more care. We make arrangements for hospice and after that cremation and burial. Often she is not sure who I am. "It's Clifton," I say, "your son." I repeat a succession of statements each time we speak, hoping they will have some kind of mnemonic power: I am married, have kids who are galloping into adulthood, teach in Atlanta. I ask about the weather. "Thank you for calling," she mumbles, and then after a few minutes puts the phone down. She is forgetting who I am. She is forgetting everything.

I visit Mom the morning after Susan's memorial service. We had placed my sister's ashes next to where I surreptitiously buried my father. Afterwards we all gathered at

Sabrina's house for a lunch of po'boys from Domilise's. We talked about Mom and her declining health and signed forms anticipating the inevitable. We all knew we would be mourning again soon, though I wondered if by this point we were all immune. We've been grieving our entire lives.

Over the course of a few days, children and grandchildren took the elevator up to Mom's apartment. Marie visited and was shaken by what she saw.

"I think she's waiting to see you," Bill told me, waiting to see her children one last time.

So I visit. I bring her a blueberry muffin. Kinta is there, dutiful as always. "Mom, you want a cup of coffee with your muffin?" Mom lays crumpled up in bed, barely able to raise her head. She drinks a few sips, eats some crumbs, and says few words, but she keeps closing her eyes and drifting away.

My children are there and my wife, and we all take turns trying to make conversation. Mom seems exhausted. Pain ripples across her lips. Her legs shake as if she is cold or has a fever. I am not sure she knows who we are, though for a second I see a flash of recognition in her eyes.

We visit for an hour then prepare to leave. It is the last time I will speak to my mother. The questions I couldn't ask, the things I could never say. All of it will remain unsaid.

ACKNOWLEDGMENTS

I COULD NOT HAVE WRITTEN THIS BOOK WITHOUT THE help of my family, especially my mother. It was not easy for her; I know she holds within her many regrets and considerable bitterness. Nonetheless, she supported my research by agreeing to be interviewed and allowing me access to sensitive mental health records. My sisters were also helpful (even if Susan refused to be interviewed), especially Kinta who read through an early draft and occasionally forwarded photographs and other records.

Anna Von Veh, Sita Ranchod-Nilsson, Ulf Nilsson, Elizabeth Gallu, Thom McClendon, and Kerry Ward also read early drafts and offered helpful suggestions. Emory University colleagues Angelika Bammer, Anna Grimshaw, Howard Kushner, Rosemarie Garland-Thomson, and Abdul Jan Mohammed read the early chapters. I am particularly thankful to Angelika and Anna for inviting me to share two chapters with their graduate seminar. Undergraduate students in my "Writing Memory" course made many helpful observations, as did audiences at various readings.

Three decades ago, my undergraduate mentor, Gaby Spiegel, was the person largely responsible for getting me into the graduate program at Johns Hopkins University. Gaby read a later draft. Her comments and wisdom helped me complete *History Lessons*.

Librarians and archivists were unstintingly helpful. I am also thankful to Drs. Goldberg, Chance, and Giustra. David Raney offered expert editorial advice as did Chantal Clarke. I am especially indebted to my agent, Jessica Papin, and my editor, Dan Crissman, for making everything happen.

Pamela Scully has been there, always. Spouse, colleague, muse, Pamela supported the project when all I had was a few vague questions. This book is for her, with love.

0000122682883